THE SUSTAINABILITY PRACTITIONER'S GUIDE TO MULTI-REGIONAL INPUT-OUTPUT ANALYSIS

JOY MURRAY, EDITOR
MANFRED LENZEN, EDITOR

THE SUSTAINABILITY PRACTITIONER'S GUIDE TO MULTI-REGIONAL INPUT-OUTPUT ANALYSIS

JOY MURRAY, EDITOR
MANFRED LENZEN, EDITOR

COMMON
GROUND

First published in 2013 in Champaign, Illinois, USA
by Common Ground Publishing LLC
as part of the On Sustainability Book Series

Library of Congress Cataloging-in-Publication Data

The sustainability practitioner's guide to multi-regional input-output analysis / Joy Murray, editor, Manfred Lenzen, editor.
    pages cm
 ISBN 978-1-61229-190-1 (pbk : alk. paper) -- ISBN 978-1-61229-191-8 (pdf)
 1. Sustainability. 2. Environmental responsibility. 3. Sustainable development. 4. Input-output analysis. I. Murray, Joy. II. Lenzen, Manfred.

 GE195.S8725 2013
 339.2'3--dc23

                          2013007618

Cover Image: John Murray, 2013

This book is dedicated to Graham Treloar, an outstanding input-output researcher and treasured colleague and friend, who sadly died in 2008. If he had been alive today he would not only have contributed a chapter to this book but also would have been an enthusiastic supporter of the whole project.

Table of Contents

Foreword xiii

Preface.. xv

Part I: Introducing Multi-Regional Input-Output Analysis

Chapter 1: What is MRIO: Benefits and Limitations.......................... 1
Keiichiro Kanemoto and Joy Murray

Chapter 2: Uncertainty and Variability in MRIO Analysis 10
Anne Owen

Part II: World MRIO Frameworks

**Chapter 3: The Global Trade Analysis Project and the GTAP
Database** 21
Terrie L. Walmsley, Angel H. Aguiar, and Badri Narayanan

**Chapter 4: Transnational Interregional Input-Output Tables: An
Alternative Approach to MRIO?** 33
Satoshi Inomata and Bo Meng

Chapter 5: The World Input-Output Tables in the WIOD Database 43
Erik Dietzenbacher, Bart Los, and Marcel Timmer

**Chapter 6: EXIOBASE – A Detailed Multi-Regional Supply and Use
Table with Environmental Extensions** 52
Arnold Tukker

Chapter 7: The Eora MRIO 63
Dan Moran

**Chapter 8: Simplification of Multi-Regional Input-Output Structure
with a Global System Boundary: Global Link Input-Output Model
(GLIO)**.. 69
Keisuke Nansai, Shigemi Kagawa, Yasushi Kondo, and Sangwon Suh

Chapter 9: The Global Resource Accounting Model (GRAM) 79
Kirsten S. Wiebe, Christian Lutz, Martin Bruckner, and Stefan Giljum

Part III: Special MRIO Frameworks

Chapter 10: Distribution of CO_2 Emissions in China's Supply Chains: A Sub-national MRIO Analysis...89
Kuishuang Feng and Klaus Hubacek

Chapter 11: Waste Flows in Multi-regional Input-Output Models....95
Christian John Reynolds and John Boland

Chapter 12: An Enterprise MRIO for a University.........................104
Christopher Dey, Sandra Harrison, Manfred Lenzen, and Joy Murray

Chapter 13: Water Footprints for Spanish Regions Based on a Multi-Regional Input-Output (MRIO) Model 113
Ignacio Cazcarro, Rosa Duarte, and Julio Sánchez-Chóliz

Part IV: Case Studies

Chapter 14: Case Study Using the GTAP Database: Footprint and Supply Chain Analysis of Dutch Consumption...................................125
Harry Wilting

Chapter 15: Consumption-Based Inventory of Global Land Use: An Application of the GTAP Database 136
Yang Yu, Kuishuang Feng, and Klaus Hubacek

Chapter 16: Accounting for Environmental Responsibility: A Case of Asian Countries Using the Asian International Input-Output Model 140
Xin Zhou and Alexandra Marques

Chapter 17: Sustainability Assessment from a Global Perspective with the EXIOPOL Database 157
Richard Wood and Kjartan Steen-Olsen

Chapter 18: Structural Decomposition Analysis of the Energy Consumption in China and Russia — An Application of the Eora MRIO Database 170
Jun Lan and Arunima Malik

Chapter 19: Estimating Global Environmental Impacts of Goods and Services Produced in Japan Using a Global Link Input–Output Model (GLIO) 180
Keisuke Nansai, Shigemi Kagawa, Yasushi Kondo, Susumu Tohno, and Sangwon Suh

Chapter 20: Carbon Emissions and Materials Embodied in Emerging Economies' Trade: An Application of the Global Resource Accounting Model (GRAM) 190
Martin Bruckner, Leisa Burrell, Stefan Giljum, Christian Lutz, and Kirsten S. Wiebe

Part V: The Role of MRIO in Global Governance

Chapter 21: Policy Discussions Using Inter-Country Input-Output (ICIO) Systems 203
Norihiko Yamano and Colin Webb

Chapter 22: The Use of MRIO for multilateral trade policy 211
Christophe Degain, Hubert Escaith, and Andreas Maurer

Chapter 23: The Importance of Input-Output Data for the Regional Integration and Sustainable Development of the European Union . 220
Joerg Beutel, Isabelle Rémond-Tiedrez, and José M. Rueda-Cantuche

Chapter 24: How MRIO Can Help APEC to Address its Environmental Objectives 240
Kirill Muradov and Akhmad Bayhaqi

Chapter 25: Environmental-Extended Input-Output Analysis in the System of Environmental-Economic Accounting (SEEA) 250
Julian Chow

Part VI: The Future

Chapter 26: Current and Future Policy Applications of MRIO Research 267
Thomas Wiedmann and John Barrett

List of Contributors 276

Index 279

Icon of the sea
Oil on canvas, 90x100cm
Dagmar Hoffmann

Foreword

When Joy and Manfred asked me to try my hand at writing the Foreword for this cooperative effort they have so enthusiastically edited, I was both thrilled and puzzled. Actually, there is nothing more difficult for a statistician, accustomed to letting numbers and equations do the talking, than using plain words. Especially writing for the wider public, because it is for them that this project was implemented. This book aims at explaining in non-technical terms how we can make sense of one of the most complex and far reaching movements of globalisation: the internationalisation of production networks. And when I say making sense, I refer to what we as statisticians consider is the proper way of making sense of things: by measuring them.

The internationalisation of production, often referred to as the global supply chain, is one of the major economic, social and environmental phenomena of the early 21st century. It has changed in 20 years the lives of billions of people. The opening of China, India and Russia added a huge potential to world productive capacities, dramatically changing the supply-demand balances of natural resources and labour global markets. In the process millions of people have been lifted out of poverty in some parts of the developing world, while a new middle class has emerged in developing countries, nearly doubling the field of play for internationalisation. But the consequences of economic globalisation can also be negative, even for those developing countries that have been the main beneficiaries. *Compressed development*, as the rapid pace of industrialisation is known, creates new policy challenges that national governments are still not well equipped to deal with, especially in the social and environmental dimensions. In the old industrialised world the rise of those emerging countries is often seen as a challenge to job security and social mobility.

But the statistical resources to monitor internationalisation and analyse its effects are still far from perfect. As with many social constructs, Statistics often changes in response to a crisis. After the 1929 depression new statistical efforts were geared towards developing a better understanding of the domestic economy, creating the System of National Accounts. Since the 2008-2009 global crisis we know that we need to go further and create a System of International Accounts. But it is indeed extremely difficult to map the complexity of supply routes through the vast economic system of global supply chains, which inter-connects numerous industries in many countries. The challenge is to track their production on the world map from the first productive tasks down to their final market of consumption while keeping a record, in the process, of the various economic, social and environmental implications.

Yet there is no option but to tackle this enormous task. When the proper statistical information is not available to support policies, responses are sometimes based on misperceptions and policies can become counter-productive. Many researchers are working today on modernising the statistical system in order to provide the appropriate information that will support evidence-based policy; many of them have participated in the elaboration of this book.

Thus, this is a very special book: written by leading experts for the non-experts. The result is a curious product that has two personalities. Some books are windows, others are mirrors; this one is both a window and a mirror at the same time. If you are an expert you will use it as a mirror in order to reflect on your subject. My suggestion for you, the expert: read it from front to end. If you are not an expert and want to use it as a window to discover a new and fascinating subject, my suggestion is to read it from end to front, as a Manga: You will first understand *"why"* this effort was made and *"what"* analytical and policy recommendations can be derived from the results. Then you will move upstream, as we say in our jargon and discover *"how"* the information required to reach these results was built.

In both cases you will learn a lot in the process.

Hubert Escaith,
Chief Statistician,
Economic Research and Statistics Division,
World Trade Organization

Preface

Come here everyone. We would like to invite you who are reading this book to come here with us now, come, sit ...and look across the peaceful waters of Port Bradshaw. ...we will sit under these beautiful trees and listen to them whisper to us. We will sit together, share some food and talk about sustainability. ...We've been here a long time, the longest time, longer than any other culture. Our culture is strong ...We'll share with you about how we relate to Country – to the land, the animals, the winds, the spirit beings, the water, the rocks and everything around us. Animals and plants, seasons and winds, water and ceremony, stars and shells, these are all sentient, they all have knowledge, they are all connected to us. We are not separate and we are not above them. ...Yolŋu people look at things in a connected way. The land and the nature are all kin. We don't separate the environment as something different from us. It's not a 'thing'. We understand that everything is our kin, everything is connected. ...Everything has a rightful place in the world. (Burarrwanga, L., Ganambarr, M., Ganambarr, B., Suchet-Pearson, S., Lloyd, K. & Wright, S (2012). Learning from Indigenous Conceptions of a Connected World. In J. Murray, G. Cawthorne, C. Dey & C. Andrew (Eds) (2012) *Enough for All Forever*. Common Ground Publishing LLC, Champaign, Illinois, p.3.)

Everything is connected. …Everything has a rightful place in the world…

…But the edges of those 'rightful places in the world' are beginning to blur. More and more things are being shifted around. We no longer know where the rightful place is – From where did this come? Where does it belong? Who brought it here? Why? Where is it going?

These might be philosophical questions about power, belonging and transformation. They might also be practical questions about ownership, travel and toil. For as things travel the world they shape-shift too; are transformed, constituted, consumed, used and reconstituted. From all of this activity there remains an intricate and indelible trail.

The chapters in this book map that indelible trail. They do so with the spirit of mapmakers throughout history; building on what has gone before, paying attention to detail, gathering data and drawing or re-drawing boundaries. All of it painstaking work knowing that these maps are important for survival.

Perhaps that seems like a dramatic claim for a book about multi-regional input-output analysis, but what other quantitative tools do we have that can trace circuitous supply routes through a vast global economic system of ever-growing complexity, connect numerous points of production on the world map with a comprehensive compendium of environmental interventions, yield information about trade-offs between environmental and economic objectives, and thus guide policy decisions on global matters like allocating fair shares of the climate change burden? There is nothing else with the power and elegance of multi-regional input-output analysis; hence this book.

We want to bring this power and elegance to the toolboxes and fingertips of policy makers everywhere. To do this we have invited experts from around the world to describe in non-technical language their tools, how and for what purpose they can be used and the new dimensions that they bring to policy making.
The book is arranged in six parts. The first part is comprised of an introduction to MRIO analysis for the non-expert reader by Keiichiro Kanemoto and Joy Murray, who also discuss the benefits and limitations of MRIO-based analytical techniques. This is followed by a chapter, written by Anne Owen, which focuses on uncertainties in MRIO frameworks.

Part two describes in non-technical language seven major tools that have recently been constructed. These are the MRIO databases that map the movement of goods and services around the world. The Global Trade Analysis Project (GTAP) is introduced by Terrie Walmsley, Angel Aguiar and Badri Narayanan. The Institute of Developing Economies – Japan External Trade Organisation (IDE-JETRO) database is explained by Satoshi Inomata and Bo Meng. Erik Dietzenbacher, Bart Los, and Marcel Timmer write about the World Input-output Database (WIOD). EXIOBASE, a global multi-regional environmentally extended supply and use / input-output database, is described by Arnold Tukker. Daniel Moran introduces us to the Eora database, named after the Aboriginal people who are the traditional custodians of the land on which the University of Sydney is built. The Global Link Input-output model (GLIO) is presented by Keisuke Nansai, Shigemi Kagawa and Yasushi Kondo. Kirsten Wiebe, Christian Lutz, Martin Bruckner and Stefan Giljum describe the Global Resource Accounting Model (GRAM). Each database has its unique features. All are built to perform specific tasks for furthering our knowledge of how extensive and intricate supply chains affect the lives of millions of people one way or another.

Part three provides some special sub-national MRIO frameworks designed to track the movement for example of carbon, water and waste through national economies. Kuishuang Feng and Klaus Hubacek illustrate how China's regions are characterised by substantial discrepancies in CO_2 emissions. Christian Reynolds and John Boland operationalize an Australian MRIO model tracing waste flows throughout the nation. Christopher Dey, Sandra Harrison, Manfred Lenzen and Joy Murray take a multi-regional approach to accounting for financial and environmental flows throughout a large organisation – a university – that is treated as though it were a town. Finally, Ignacio Cazcarro, Rosa Duarte and Julio

Sánchez-Chóliz show how considerable climatic variation can lead to pronounced differences in water supply issues across Spain.

Having described the tools, Part four shows them in action, providing case studies of how they can be used by countries and organisations.

It begins with two case studies using the GTAP database. Harry Wilting provides a footprint and supply chain analysis of Dutch consumption while Yang Yu, Kuishuang Feng, and Klaus Hubacek present a consumption-based inventory of global land use. They track land use along global supply chains to examine how much and for what purpose one country has been acquiring land in other countries' territories in order to support the consumption and lifestyles of their own citizens. Xin Zhou and Alexandra Marques illustrate an application of the IDE-JETRO Asian International Input-Output Model (AIO) in assigning environmental responsibility across countries and compare national inventories based on producer responsibility, consumer responsibility, income responsibility and shared environmental responsibility. In the next chapter Richard Wood and Kjartan Steen-Olsen give a global perspective of the issues of economic development, employment and greenhouse gas emissions, applying the EXIOPOL database to an analysis of consumption by product in the European Union. Jun Lan and Arunima Malik illustrate for the example of the Eora database how MRIO frameworks can be used for Structural Decomposition Analysis, a technique that can reveal underlying drivers of environmental change.

In the next chapter Keisuke Nansai, Shigemi Kagawa, Yasushi Kondo and Susumu Tohno illustrate the use of the GLIO for estimating the global impacts of goods produced and services provided in Japan. Finally in this section Martin Bruckner, Leisa Burrell, Stefan Giljum, Christian Lutz and Kirsten Wiebe show how the global resource accounting model (GRAM) can shed light on emissions and material metabolism of emerging economies.

Part five provides insights into MRIO's role in global governance with contributions from five major world institutions. Norihiko Yamano and Colin Webb describe how the OECD use inter-country input-output (ICIO) systems for facilitating policy discussions; in particular looking at the OECD's use of MRIO to examine changes in global trade structure. Christophe Degain, Hubert Escaith and Andreas Maurer discuss how the measure of *trade in value added* helps us to understand the economic and social dimensions of trade and provides the World Trade Organisation (WTO) with new perspectives on the design of trade policies and the analysis of their impact. Next Joerg Beutel, Isabelle Rémond-Tiedrez and José Manuel Rueda-Cantuche describe a European Union perspective on the importance of input-output data for regional integration and sustainable development. They describe the process of developing aggregate supply and use and input-output tables and environmentally extended input-output tables and show how the new input-output database can be used for a more profound assessment of Structural Fund interventions and regional policies than has been possible in the past.

Kirill Muradov and Akhmad Bayhaqi demonstrate how MRIO can help APEC to address its environmental policy objectives, especially the Green Growth initiative, and improve overall policy making. Finally Julian Chow introduces the Environmental-Extended Input-Output Analysis in the System of

Environmental-Economic Accounting (SEEA) of the United Nations Statistical Division. He illustrates a number of techniques that may be applied to data from these SEEA-compliant EE-IOTs to answer policy questions.

Part six concludes the book with an insightful chapter on the future of MRIO research by Thomas Wiedmann and John Barrett.

This book's purpose is to introduce non-experts to what is a highly sophisticated, technical, multi-disciplinary and, until recently, almost exclusively academic area of study. However it is one that has now caught the attention of decision-makers involved in crafting policies on globally important environmental, social, and economic issues. The book has been written to bridge the gap between the academic sphere and the practical – describing the new tools in as non-technical terms as possible and illustrating the application of these tools to some of the urgent problems that are confronting us today.

The artworks that introduce each section are by Dagmar Hoffmann, a Czechoslovakian born, Australian artist who has exhibited widely in Sydney and the Czech Republic and been awarded prizes in Australia and Japan (http://www.dagmarhoffmann.com.au/dagmarpaintings.html).

Joy Murray
Manfred Lenzen
Sydney, January 2013

Part I: Introducing Multi-Regional Input-Output Analysis

Bird guardians
Oil on canvas 111 x 143cm
Dagmar Hoffmann

Chapter 1: What is MRIO: Benefits and Limitations

Keiichiro Kanemoto and Joy Murray

The expansion of globalization over the last few decades has generated economic growth. International trade has grown rapidly compared, for example, to GDP (gross domestic product), population, and CO_2 emissions (Fig. 1). Exports of goods and services are about 49 times larger today than they were in 1970. As a result of globalization we have changed our economic, social, and environmental perspectives from the local to the global level. Expansion of international trade has changed our production and consumption patterns completely, and has generated wide-ranging side effects. For example, in 2008 Westphal and colleagues found the greater the degree of international trade, the higher the number of invasive alien species[1].

Another recent interest for consumers, researchers, and companies is supply chain and life cycle thinking. This supply chain and life cycle perspective has shifted our focus from the goods themselves to the processes involved in developing and transporting those goods. For example, food companies have started to show not only the calories and ingredients but also the food mileage and genetic-modification on the label of foods. Numerous studies on bottom-up life cycle assessment (LCA) that accumulate the emissions of each individual supply chain, have been conducted on various products[2]. But bottom-up LCA cannot

---

[1] Westphal, M. I., Browne, M., MacKinnon, K., & Noble, I. (2008). The link between international trade and the global distribution of invasive alien species. *Biological Invasions*, *10*(4), 391-398.

[2] Finnveden, G., Hauschild, M. Z., Ekvall, T., Guinée, J., Heijungs, R., Hellweg, S., Koehler, A., et al. (2009). Recent developments in Life Cycle Assessment. *Journal of environmental management*, *91*(1), 1-21. Roy, P., Nei, D., Orikasa, T., Xu, Q., Okadome, H., Nakamura, N., & Shiina, T. (2009). A review of life cycle assessment (LCA) on some food products. *Journal of Food Engineering*, *90*(1), 1-10.

cover the whole economy and all production stages because of limitations of resources and data[3]. Therefore an input-output approach, analyzing supplier and demander interdependencies right along the production chain is now being used as a tool of top-down LCA that covers emissions of the whole supply chain and economy (for more information see Further readings: *The Sustainability Practitioner's Guide to Input-Output Analysis*).

Figure 1. The growth of world exports of goods and services, GDP, population, and CO_2 emission from 1970 to 2010 (1970 = 1)[4].

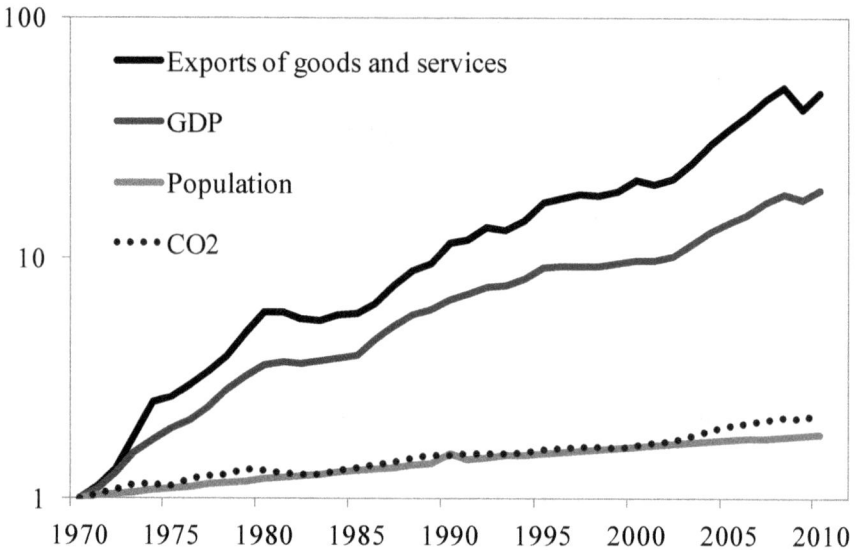

Against the background of the expansion of globalization and recent interest in a life cycle perspective, multi-region input-output (MRIO) analysis is emerging as a way to analyse the global supply chain[5]. There are several initiatives underway to construct global MRIO tables (see Section 2), and studies using MRIO as a

[3] Lenzen, M. (2001). Errors in Conventional and Input-Output-based Life-Cycle Inventories. *Journal of Industrial Ecology*, 4(4), 127-148. Lenzen, M., & Treloar, G. (2006). Differential Convergence of Life-Cycle Inventories toward Upstream Production Layers Implications for Life-Cycle Assessment. *Journal of Industrial Ecology*, 6(3), 137-160.

[4] Friedlingstein, P., Houghton, R. A., Marland, G., Hackler, J., Boden, T., Conway, T. J., Canadell, J. G., et al. (2010). Update on CO2 emissions. *Nature Geoscience*, 3(12), 811-812. Nature Publishing Group. Boden, T., Marland, G., & Andres., R. J. (2011). Global, Regional, and National Fossil-Fuel CO2 Emissions. Carbon Dioxide Information Analysis Center, Oak Ridge National Laboratory. *U.S. Department of Energy, Oak Ridge, Tenn., U.S.A.* United Nations. (2012). National Accounts Main Aggregates Database. Retrieved from http://unstats.un.org/unsd/snaama/

[5] Wiedmann, T. (2009). A review of recent multi-region input–output models used for consumption-based emission and resource accounting. *Ecological Economics*, 69(2), 211-222. Sato, M. (2012). Embodied carbon in trade: a survey of the empirical literature. *Centre for Climate Change Economics and Policy Working Paper No. 89.*

tool can be found in many top academic journals[6]. These studies have mainly focused on global carbon footprints and consumption-based national CO_2 emissions but there are many other environmental and social indicators that can be explored using global MRIO as well as exploring other scales of MRIO such as multi-regions within a country (see Sections 3 and 6). The applications of MRIO are increasing, which in turn will increase the relevance of MRIO to global governance (see Sections 4 and 5).

In this chapter, we introduce the concept and a short history followed by some of the benefits and limitations of MRIO.

Concept and History of MRIO

The single-region input-output (SRIO) table that is published by central and local governments includes one region's intermediate demand (inter-industry transactions), final demand (final consumption of households etc.), exports, imports, and value added across a certain number of industries and commodities. The SRIO model allows us to analyse only the supply chains within one region, because single region IO tables do not record what happens to exported goods or services once they are exported or what occurs in the making of imported goods or services before they are imported. In MRIO tables on the other hand, exported goods and services from one region are treated as inputs to industries in other regions, and imported goods and services of one region pass through other regions' production processes. Therefore, we can trace the supply chain between regions. MRIO analysis, footprint analysis, and interregional LCA share the common aim of tracing supply chains, thus MRIO analysis is seen as a promising approach to apply to environmental problems.

In Fig. 2 we show an example of a two-regional input-output table with two sectors: goods and services. The highlighted column means Region 1's Service sector buys 10 units of Goods and 20 units of Services from Region 1, imports 10 units of Services from Region 2, and pays 5 units to value added for the provision of Services. The highlighted row means Region 2 exports 30 units of Goods to the Goods sector and 15 units to final demand of Region 1; and sells 20 units of Goods to the Goods sector and 10 units to final demand of Region 2. Therefore

---

[6] Davis, S. J., & Caldeira, K. (2010). Consumption-based accounting of CO2 emissions. *Proceedings of the National Academy of Sciences of the United States of America*, *107*(12), 5687-5692. Davis, S. J., Peters, G. P., & Caldeira, K. (2011). The supply chain of CO2 emissions. *Proceedings of the National Academy of Sciences of the United States of America*, *108*(45), 18554-9. Lenzen, M., Moran, D. D., Kanemoto, K., Foran, B., Lobefaro, L., & Geschke, A. (2012). International trade drives biodiversity threats in developing nations. *Nature*, accepted. Peters, G. P., Marland, G., Le Quéré, C., Boden, T., Canadell, J. G., & Raupach, M. R. (2011). Rapid growth in CO2 emissions after the 2008–2009 global financial crisis. *Nature Climate Change*, *2*(1), 2-4. Peters, G. P., Minx, J. C., Weber, C. L., & Edenhofer, O. (2011). Growth in emission transfers via international trade from 1990 to 2008. *Proceedings of the National Academy of Sciences of the United States of America*, *108*(21), 8903-8. Steinberger, J. K., Roberts, J. T., Peters, G. P., & Baiocchi, G. (2012). Pathways of human development and carbon emissions embodied in trade. *Nature Climate Change*, *2*(2), 1-5. Nature Publishing Group.

the total export of Goods from Region 2 to Region 1 is 45 units. From this simple example it can be seen that in essence MRIO tables describe economic structure, inter-industry and inter-regional transactions.

Figure 2. Example of a two-regional input-output table

		Region 1		Region 2		Final demand	
		Goods	Services	Goods	Services	Region 1	Region 2
Region 1	Goods	100	10	5	0	30	0
	Services	0	20	0	0	20	5
Region 2	Goods	30	0	20	0	15	10
	Services	10	10	0	5	0	10
Value added		5	5	50	30		

Constructing an MRIO table is not a complex task, but it is data-hungry and computationally intensive. Therefore, global-scale MRIO tables were not constructed until around 2008. MRIO tables consist of domestic and interregional trade blocks (Fig. 3). SRIO tables are used to build the domestic trade blocks of the MRIO table for each region. National Statistical Offices regularly conduct industry surveys to compile SRIO tables of their own country. However, official statistics on the interaction of regional trade blocks often don't exist so several ways to estimate interregional trade blocks in MRIO tables have been developed. One way is to survey industries directly to find out which industries or sectors of the economy (which could be households) use imported commodities from which countries. For example the UK might import cars from Japan and Germany. However unless we actually survey to find out who buys which cars we cannot know where the German manufactured cars end up and where the Japanese ones go. Surveying increases the accuracy in tracing supply chains. It is particularly important where there are significant differences in the way the product is manufactured in different countries. Another way to estimate interregional trade blocks in MRIO tables is known as the non-survey method, or trade coefficient approach. It is a way of constructing interregional trade blocks without surveying to find out which industries use imported commodities from which countries. In the non-survey method, import tables are disaggregated into a certain number of trade blocks using bilateral trade statistics. However, to continue the example above, this method will not tell you exactly who buys which cars. It will tell you what percent of imported cars come from Germany and what percent come from Japan. These will then be lumped together as *imported cars* and pro-rated according to the percentage of imported cars bought by households, government or various other sectors of the economy irrespective of where they originated.

Whichever method is chosen matrix balancing should always be conducted on MRIO tables to fulfil the general economic theory that gross input, such as raw materials and wages, is equal in monetary value to gross output such as household consumption. This is the same process as we apply to the construction of SRIO tables.

Figure 3. Construction process of MRIO table with non-survey method; the MRIO table is constructed from SRIO tables, import tables, and bilateral trade data. SRIO tables are aligned as diagonal domestic blocks and value added blocks in the MRIO table (see left side). In the non-survey method, intermediate demand (ID) and final demand (FD) in import tables are disaggregated into bilateral import tables using bilateral trade data (region (R) x -> region (R) 1) (see right side).

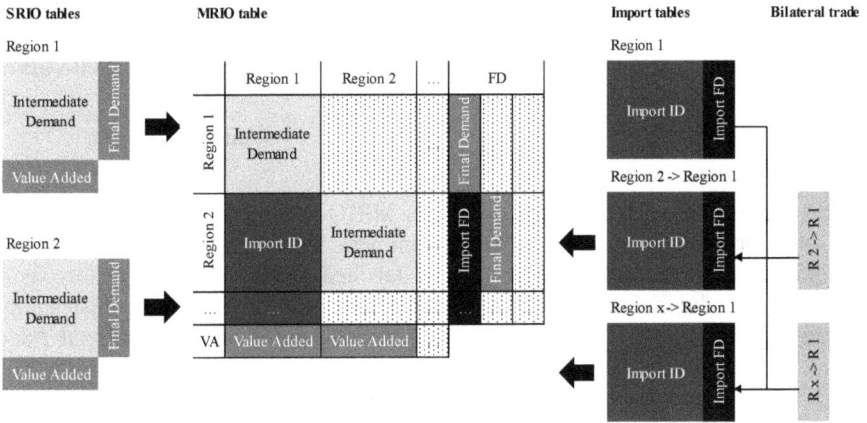

MRIO or IRIO

There is a lack of consensus about the terminology for multi-/inter-national input-output modelling: should MRIO or IRIO be used as the correct term for this type of model? According to definitions from the literature, IRIO (inter-regional input-output) tables have 'perfect' trade matrices based on real information gained from surveying industries, whereas MRIO tables have derived trade matrices based on trade coefficients (as described above), because they do not conduct surveys and therefore have incomplete information[7].

Both, IRIO and MRIO are aiming at a structure where bilateral trade is separated by industry; that is, both aim to understand what a country imports, what kind of goods are imported, and who or which industry uses these goods. However the means to achieve this aim may be different. Some researchers go beyond the trade coefficient (non-survey) approach for compiling MRIOs by incorporating superior data, e.g. industry-specific import reports (Fig.4). The aim is to approximate IRIO data.

---

[7] Guo, D., C. Webb, and N. Yamano. 2009. *Towards harmonised bilateral trade data for inter-country input-output analyses: statistical issues*. STI Working Paper 2009/4 (DSTI/DOC(2009)4). Paris, France: Organisation for Economic Co-operation and Development (OECD).

Figure 4. Transition between MRIO and IRIO; as we use or survey actual international trade blocks, the expression changes. MRIO starts with the non-survey method with a fixed proportion of import and domestic tables. The expression of MRIO is gradually changed to IRIO if we use surveyed import tables and surveyed international trade blocks. Once we have surveyed all international trade blocks, then we call the table an IRIO table.

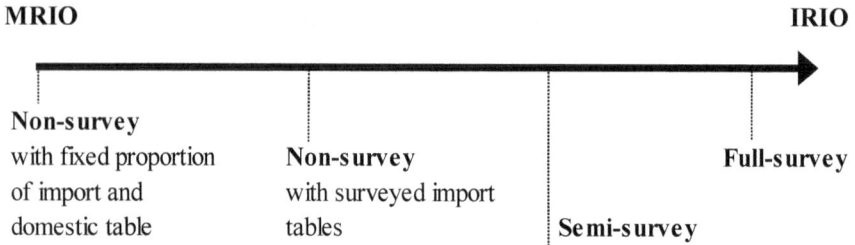

MRIO   **IRIO**

Non-survey
with fixed proportion   **Non-survey**   **Full-survey**
of import and   with surveyed import
domestic table   tables   **Semi-survey**

This means that whether a table is IRIO or MRIO is a matter of quality as well as of structure, and it is not clear what minimum level of quality or real data is required before an MRIO table can be labelled IRIO. Adopting quality as a criterion in the definition makes it subjective and is potentially misleading. Strictly speaking, IO tables comprising several nations and international trade data will always be MRIO tables because the information necessary for a true IRIO table is not likely to become available in the foreseeable future. Therefore, it is suggested that the term IRIO should only be used if a clear and significant superiority of data quality compared to MRIO tables can be demonstrated. In this chapter, and also throughout the book, the term 'MRIO' is used to denote the entire continuum.

History

Since Walter Isard proposed the initial framework of MRIO tables in 1951, there has been a history of MRIO analysis. From the start, researchers recognised the MRIO table as a good tool with which to understand the world economic structure; and since then have worked to develop an ever more sophisticated form of the table that could be applied to the new world economic framework. Although not part of mainstream macroeconomics, MRIO, for example, has long been one of the favoured tools of regional science for examining the regional impacts of national or global processes of economic and social change. The development of regional science has facilitated compilation of regional input-output tables such as in China and Japan and has helped clarify structural change and regional interdependence. Since about 2005 MRIO frameworks have attracted attention from researchers investigating environmental issues. The growth in international trade has also prompted researchers in other fields to take interest in the use of MRIO tables. For example, there is a growing interest in value added trade (Chapters 21, 22, 23). The volume of trade from developing to developed countries continues to grow rapidly, but people tend to think that developing countries don't add a lot of value because their main role is in assembling. Recent studies have found that the MRIO table is a useful way to estimate which country and sector adds actual value, i.e., value added trade.

MRIO Benefits and Limitations

MRIO analysis is becoming recognized as a powerful tool to analyse environmental problems because there is no other way to analyse complex global supply chains. Here we introduce three benefits: coverage, extension, and application capacities; and two limitations: sector detail and time lag.

Benefits

The first benefit of MRIO is geographical and supply chain coverage. MRIO tables cover the whole economic structure of multi-regions and exports and imports within and outside these regions as well as the long and complex supply chains. For example, crops consumed by US households are imported from China, Chinese companies import agricultural machinery from Japan, Japanese companies import tyres from Malaysia and so on in a never ending chain of transactions around the globe. In MRIO analysis, we can avoid having to set arbitrary geographical and production stage boundaries, which is something that we have to be concerned with when using bottom-up LCA and footprinting.

The second beneficial feature of MRIO is that we can easily extend MRIO analysis to any company. For example, despite the growing interest in carbon footprints of products (CFP), estimation of CFP is hard work for most companies. This is because usually companies need to trace long supply chains of many individual products. MRIO can play the role of tracer of products. Thus we only need to prepare the MRIO table and direct carbon emissions by sector to estimate CFP. In addition to its tracer role, we can use MRIO results in a screening and complementary role with existing bottom-up LCA results[8].

The third example of a benefit of MRIO analysis is its application capacities. MRIO uses the same framework as SRIO, so we can use many of the techniques that we have already used in SRIO analysis such as structural path analysis (SPA)[9] and structural decomposition analysis (SDA)[10]. Structural path analysis (SPA) traces the supply chains of production processes, so using SPA we can find, for example, the path by which Australia exports iron ore to Thailand, Thailand manufactures automobiles using the iron ore and Japanese households 'consume' the automobiles.

---

[8] Huang, Y. A., Lenzen, M., Weber, C. L., Murray, J., & Matthews, H. S. (2009). The Role of Input–Output Analysis for the Screening of Corporate Carbon Footprints. *Economic Systems Research*, *21*(3), 217-242. Suh, S., & Huppes, G. (2002). Missing inventory estimation tool using extended input-output analysis. *The International Journal of Life Cycle Assessment*, *7*(3), 134-140.

[9] Peters, G. P., & Hertwich, E. G. (2006). Structural analysis of international trade: Environmental impacts of Norway. *Economic Systems Research*, *18*(2), 155-181.

[10] Baiocchi, G., & Minx, J. C. (2010). Understanding Changes in the UK's CO2 Emissions: A Global Perspective. *Environmental Science & Technology*, *44*(4), 1177-1184.

SDA is a technique that can be used to find the driving factors behind changes in emissions over time. SDA can distinguish driving factors (e.g. emission intensities, economic structure, consumption categories, consumed amount).

Limitations

One of the limitations of MRIO and perhaps the biggest reason why input-output analysis isn't widely used for product LCA is lack of sector detail. Because of low sector resolution input-output analysis sometimes can't distinguish important differences between products or industries. Sometimes one region provides different sector classification tables from another region. Because of this most MRIO studies aggregate input-output tables into one common classification (see Section 2). That process loses quite a lot of information and makes product LCA more difficult. One solution against this limitation is a hybrid-LCA approach. A hybrid approach integrates the sector- (input-output approach) and process-based (bottom-up LCA approach) data[11]. However, the integration of MRIO tables and process-based data presents formidable challenges.

Another limitation of MRIO is time lag. Benchmark input-output tables are usually published every five years and released a few years later because compilation of these tables needs a considerable amount of data and labour. Of course some countries publish annual input-output tables such as Japan and USA, but these tables are aggregated or simplified based on benchmark input-output tables. Because the MRIO table is an assembly of SRIO tables, differences in the publication year of SRIO tables affects the quality of the MRIO.

The following sections of the book illustrate further benefits and limitations and provide examples of MRIO tables and their uses.

Further readings

Hewings, G., & Jensen, R. (1987). Regional. interregional and multiregional inout-output analysis. In P. Nijkamp, *Handbook of Regional and Urgan Economics* (Vol. 1, pp. 295-355). North Holland.

Kanemoto, K., Lenzen, M., Peters, G., Moran, D. D., & Geschke, A. (2012).

Frameworks for comparing emissions associated with production, consumption and international trade. *Environmental Science Technology* , *46* (1), 172-9.

Murray, J., & Wood, R. (2010). *The Sustainability Practitioner's Guide to Input-Output Analysis.* Urbana Champaign, Illinois: Common Ground Publishing LLC.

---

[11] Suh, S., Lenzen, M., Treloar, G. J., Hondo, H., Horvath, A., Huppes, G., Jolliet, O., et al. (2004). System boundary selection in life-cycle inventories using hybrid approaches. *Environmental Science & Technology*, *38*(3), 657–664.

Sato, M. (2012). *Embodied carbon in trade: a survey of the empirical literature. Working Paper No 89.* Centre for Climate Change Economics and Policy . Centre for Climate Change Economics and Policy .

Wiedmann, T., Wilting, H., Lenzen, M., Lutter, S., & Palm, V. (2011). Quo Vadis MRIO? Methodological, data and institutional requirements for multi-region input-output analysis. *Ecological Economics , 70* (11), 1937-1945.

Acknowledgements

The authors greatly appreciate the comments made by Manfred Lenzen (University of Sydney) and comments related to the difference between MRIO and IRIO made by Thomas Wiedmann (CSIRO), Terrie Walmsley (GTAP) and Glen Peters (CICERO).

Chapter 2: Uncertainty and Variability in MRIO Analysis

Anne Owen

Introduction

Results from multi-regional input-output (MRIO) models are increasingly being used for activities such as the calculation of consumption-based emission accounts. If these data are to be used as evidence to inform global climate policies on emissions reduction responsibilities an understanding of the robustness and accuracy of the results is required. Techniques need to be developed to calculate uncertainties in model outcomes and help communicate the findings amongst the research community and policy makers.

This chapter begins by classifying the types and causes of uncertainty in MRIO analysis, paying particular attention to uncertainty specific to MRIO systems compared to IO systems. The techniques that analysts have used to measure the uncertainty in IO and MRIO models are reviewed and the risks caused by uncertainties are briefly identified before moving on to how analysts handle risk with mitigation techniques. The chapter concludes by considering how an appreciation of uncertainty can contribute to the debate on the usefulness of MRIO models and how an improved understanding could lead to prioritisation of future model improvements.

Uncertainties are things that are not known or are partially known. In the context of uncertainty in MRIO analysis we assume the uncertainty to be quantitative in nature and that measurements of the variability in results can be made.

Uncertainty can be identified in MRIO systems in three areas:

- Uncertainty in the source data used to construct the model
- Variability and uncertainty introduced by the choice of classification and other MRIO structures

- Variability and uncertainty introduced by the choice of methods used to construct and balance the model

Uncertainty in Source Data and Processing Techniques

The ideal MRIO transactions table (MRIOT) would measure, in a consistent manner, each individual flow of money (or carbon or resource) between every industry located all over the world. In reality, no global agency is responsible for collecting these data and constructing a global MRIO system. Instead, MRIO tables are built using individual country IO tables as a starting point, together with data on international trade. This approach introduces a number of uncertainties unique to MRIO analysis that are not observed in IO approaches.

Adjustment to Common Base Currency

Firstly each country's system of national accounts reflects the differing situation within each country as to how data are collected and what data are available. An MRIOT needs to be consistent in the meaning of each category so each IO table needs; conversion to a common currency; conversion of data to basic prices; import matrices to be adjusted so that they are valued at CIF (cost, insurance and freight); import duties and commodity taxes to be excluded; negative entries dealt with; and government subsidies to be treated as 'value added' items. The choice of methods used to adjust the country level data to a common base currency can produce differences in results.

Dealing with 'Rest of World' Regions

Secondly, not every country produces an IO table, so for an MRIO to function without losing information, a 'Rest of World' (RoW) region is often created. The volume of RoW transactions can be estimated by looking at the differences between reported global trade flows and the sum of flows by countries whose data has been captured. But, alongside total transactions, a generalised structure of the economy for the ROW is also required. One approach is to pick a country that is considered representative of the RoW and multiply the structure of their IO table by RoW. For example, Malaysia is sometimes used to represent the RoW. However, for the development of the Asian International Input-Output Table (AIIOT), it is argued that the RoW region behaves similarly to the *average* Asian economy. The model creators aggregated nine Asian economies to simulate the emissions intensities, final demand and domestic IO table for the RoW. Formulating data for a RoW region clearly introduces uncertainty to the MRIO system. The Eora MRIOTs contain information for 187 countries without the use of an aggregated RoW region. Instead, Lenzen and colleagues use a generic proxy IO structure that is constructed using the detailed IO tables from the USA, Japan and Australia and is used for countries where IO tables are not produced but sectoral data on total output, value added and final demand are known.

Proportionality Assumptions

MRIOT use monetary data to represent the flows of physical commodities. This assumes that the same quantity of the commodity is represented by the same monetary amount regardless of which sector the commodity is bought by. An example of this is electricity prices varying depending on the importing industry. This assumption is further complicated in MRIO analyses where physical quantities of a good per unit of expenditure will vary considerably.

Missing Import and Export Data

The final uncertainty in source data specific to MRIOT is the absence of detailed trade flow data of intermediate demands to domestic industries. In the MRIOT structure, domestic IO tables, published by national statistical agencies are placed along the diagonal of the table. Information on imports to and exports from domestic industry is needed to populate the off-diagonal sections of the matrix. However, where many country-level IO tables provide information on sector-by-sector imports to domestic industry they lack the detail of which country the imports originate from and further data are required to completely fill the off-diagonal matrix section. In addition, exports from country A to country B should equal the figure country B reports as imports from country A. However this 'mirror statistic' rarely holds true and reasons for this mismatch in mirror statistics include:

- Valuating or recording imports and exports differently
- Different definitions as to what constitutes a trade partner
- Time lags between exports being shipped and imports received
- Different classification systems for goods and services
- Losses due to accidents in transit
- Smuggling
- Unallocated trade of confidential goods
- Trade in second-hand goods which have not required recent manufacturing or scrap and waste products for recycling or disposal
- Re-exports where goods pass through a country without any transformation and some countries record the goods' origin as the importing partner whereas others record the port the goods passed through

Re-exports are described as an increasingly problematic issue with trade data, effecting, in particular, records for China, Belgium, Hong Kong, the Netherlands, Singapore and Germany. These countries tend to have large ports.

Extension Data

Another source of uncertainty MRIO analysis shares with IO analysis is in the choice of extension data such as greenhouse gas emissions or water and resource use, applied in conjunction with the MRIOT. The latest MRIO research describes the numerous global emissions data sets available and demonstrates that the choice of emissions data input will affect the outcomes produced by the model.

Uncertainty in Sector Classification

Aggregated Sector Classifications

Common to IO analysis, the aggregation of sectors will introduce uncertainty. For example, an MRIO system may use an average multiplier for the agriculture sector in calculations of embedded water use. Within this aggregated sector, the growing of rice is more water intensive than the growing of other crops. This means that the choice of aggregation used may lead to different impact results. Grouping data together that exhibit very different energy and resource intensities will lead to calculations that use aggregated sectors containing more uncertainty than calculations that involve grouping more homogenous data.

Issues of aggregation occur when MRIO models are constructed with a sector classification that is common to each country. This is known as a harmonised MRIOT and original data from national tables are either aggregated or disaggregated to match the common classification. Summing two or more sectors to a single new sector is a simple enough procedure, but difficulties arise when a national IO entry needs to be split between two or more sectors in the new consistent sector system. Clearly additional data are needed to do this. Alongside a consistent set of country level IO tables, the additional data, such as trade, final demand, value added and extensions need to be adjusted to the common set of sectors.

Uncertainty and Variability in Model Construction Methods

Once the source data have been collected analysts make any necessary consistency adjustments and assemble them into an MRIOT. This can require the calculation of entries to populate the off diagonal sections and the application of balancing algorithms. There are numerous techniques for table construction leading to uncertainty in the final result.

Creation of the 'off-diagonal' data

As described above, the intermediate demands to domestic industry are not always known in the correct level of detail and need to be estimated in order to populate the off-diagonal portions of the MRIO table. There are numerous methods that can be used to determine these values. One such technique involves using bilateral trade data (BTD) to disaggregate the import flows. The UN's ComTrade data reveal the value of imports by sector and source country to each importing country. However, the ComTrade database does not show the sectors that use the imported goods. ComTrade data can give estimates of the import proportions but might lead to some gross assumptions being made. For example if UK industries are importing steel and ComTrade shows Mexico is the country of origin for 60% of all of the steel that is imported by the UK, then it might be assumed that for every industry in the UK, 60% of steel imported to domestic production will always come from Mexico regardless of the destination industry. This assumption is likely to introduce greater error when assessing the impacts of products from places whose domestic production is heavily reliant on imported components. In addition, once the 'off-diagonals' have been populated using an imports structure assumption it is highly unlikely that the row sums of the 'off-diagonals' will be close to reported export figures or that the condition that the row sums equal the column sums is met. Such a table is referred to as being unbalanced and is usually subjected to a balancing procedure to ensure imports equal exports conditions are satisfied.

When describing the procedure for creating the GTAP MRIO, Peters and colleagues constructed the off-diagonal matrices by considering the exports from a country first. This means working along the row of off-diagonal matrices. The authors distribute the bilateral exports from a specific country according to the import structure of each of the other n-1 countries. This means that the export balance consideration of MRIO is preserved. Due to the fact that the tables in the GTAP database are already balanced, the import structure – the sum of the off diagonal columns – is also preserved and no additional balancing is required.

Balancing the Table

When faced with an unbalanced MRIOT most analysts use the iterative technique known as RAS to bi-proportionately scale the matrix to ensure that row and column sums match. The RAS will alter values taken from the original table estimate to ensure that the table balances. Each matrix element may change as each row and column is proportionally scaled up or down until the desired row and column sums are met. Using a RAS procedure does not necessarily introduce further uncertainty to the table estimate, however post-RAS table estimates can vary because there are numerous RAS techniques used to balance the MRIOT, hence there exists variability with regard to the choice of balancing algorithm. Developments of the RAS technique include MRAS ('modified RAS'), where values that are considered to be 'perfectly known' are removed from both the matrix and the row and column sums and reintroduced at the end of the balancing process. KRAS ('Konfliktfreies RAS') allows conflicting constraints, such as row and column sums and sums of subsets of data to be introduced to the balancing calculation. Choice of balancing technique will produce different initial MRIO tables and lead to variability in MRIO results. The GRAM model (see chapter 9) does not use RAS techniques to balance MRIOTs, rather the total output and value added is adjusted to ensure row and column totals match.

Measuring Uncertainty in MRIO Systems

Errors in Source Data

Several attempts have been made to quantify standard errors of each of the input coefficients to the model taking into account the uncertainties described above, but often these data are underreported or unavailable. For example, analysts have collected standard deviations (SD) associated with the underlying source data used to make the UK IO accounts and then regressed the standard deviations across the values in the supply and use tables. Techniques for estimating the SD associated with aggregation and proportionality can be found in the further reading section at the end of this chapter.. Once estimates have been made of the individual SD associated with the steps of producing the initial transactions matrix, these can be summed together to give a total relative SD of the MRIOT. However, analysts using MRIO techniques do not usually use the MRIOT in its original format. To calculate emissions multipliers, the MRIOT is manipulated in further matrix equations. Tracing the SDs through the calculation becomes problematic because each element of the result matrix is dependent on every single other matrix element. To overcome this problem, Monte-Carlo techniques are employed.

Monte Carlo Techniques

Monte-Carlo methods involve propagating repeated random input variables through a calculation and observing the effect on the output. They have proved to be useful in estimating the SD of MRIO multipliers and work by the generation of thousands of versions of the MRIOT being created which contain random, normally or log-normally distributed adjustments to the cells of the original matrix. A matrix representing the difference between the original matrix and each of the randomly generated adjustments (MRIOT – MRIOT') has zero mean and the total relative SD of the combined input variables. Each of the thousands of newly generated tables is then subjected to the matrix calculation and the change in multipliers can be observed. Recently, Monte-Carlo techniques have been used to estimate an 89% probability that the UK's carbon footprint increased between 1994 and 2004 and to show that while uncertainties around the total Dutch carbon footprint are low, lower tiered impacts attributed at the regional and sector level contained higher uncertainty.

Monte-Carlo techniques are sometimes avoided because they can be computationally slow, but advances in computer technology and improved algorithms can improve processing times. In addition, there has been new research into alternatives to Monte-Carlo techniques that avoid the need for tens of thousands of samples by using an analytical approach to measuring error propagation. One main drawback of using these methods to understanding uncertainty in MRIO analysis is the assumption that errors in the table are normally distributed and independent of each other. The nature and construction constraints of an MRIO table mean that this is not the case. Because tables adhere to certain conditions, for example row and column sums are equal and value added equals the sum of final demand, this means that if a certain number in a particular subset is too high, the others in that group are then too low. Monte-Carlo type techniques help us to understand the propagation of error from the original source data through to the final result. However if the tables are randomly generated the analysis may not deal with systematic errors associated with the build assumptions, such as currency conversion factors, imports structures and balancing approaches made by model developers.

Structured Techniques

One method for understanding the effect of build assumptions is to build several versions of the MRIOT each with different build techniques and observe the effect on the output. For example, to estimate the magnitude of error using a domestic technology assumption (DTA) for imports analysts have investigated the difference in the consumption based emissions account of countries using the GTAP MRIOT[1] and repeated the analysis under a DTA. Here, the investigators compare 'simpler' model constructions with the results generated by the fuller more complex system to quantify uncertainty associated with the assumption. Aggregation effects can also be investigated. The difference in Nordic footprint results has been quantified first using the GTAP data with eight aggregated sectors and then the full 57 sectors. The method of currency conversion used to convert data from certain developing countries into USD has been shown to greatly affect the size of the emissions embedded in imports to the US. The analysis reveals that choosing Purchasing Price Parity[2], over Market Exchange rates increases flow sizes by a factor of two for Mexico and four for China. In addition, research into how model outcomes change when different CO_2 emissions data are used with the GTAP MRIOT concludes that much of the difference in model outcomes may be due to differences in extension data.

Monte-Carlo techniques can also be used to investigate structural effects. For example, Monte-Carlo simulations have tested the uncertainties that result upon aggregation and disaggregation. Here, the simulation repeatedly creates MRIO models and results from random numbers, then the data are disaggregated to a system with a larger number of sectors and results are produced at the disaggregated level. The table is first disaggregated using known data to give a 'true' result. Then the simulations use a mixture of known and partially known data to disaggregate the matrix and produce a result that can be compared to this 'true' value. The experiments aim to discover if the table that is disaggregated using partially known information produces results closer to the 'true' values than one that is in an aggregated state.

Managing Risk in MRIO systems

Recent reviews of MRIO models and applications have called for future MRIO analyses to inform on data reliability and numerous suggestions have been made as to how uncertainties can be measured, communicated and reduced. Recommendations include: reporting of any differences between the data used in the MRIO and the original tables; further comparisons between MRIO models to understand differences; automation of data compilation to improve timeliness of data releases and to reduce costs; and co-ordinated efforts towards MRIO data

---

[1] Andrew et al., (2009) use GTAP 6 with 87 regions and Peters and Solli, (2010) use GTAP 7.1 with 112 regions

[2] Purchasing Price Parity adjusts the prices of goods and services to represent the same volume of goods regardless of the country of purchase. It allow the relative value of currencies to be determined

provision and construction. Uncertainty and error in MRIO analysis is currently under-researched with very few studies reporting confidence intervals and even less commenting on efforts to reduce uncertainty.

Uncertainties lead to risk in model results but risk does not always have to have negative connotations. One of the outcomes of a MRIOT with some inherent uncertainty is that there is some system flexibility. The Eora system attaches standard deviation estimates to each cell value within the MRIOT giving narrowest standard deviations to the data taken from national IO tables and widest to the off diagonal values of estimated imports to intermediate demand. To balance the table the Eora model uses an optimisation routine that exploits the fact that there are uncertainties within the data. A table solution is one that satisfies the MRIOT constraints whilst using values that only deviate from the original solution by the size of the standard deviation. The Eora creators recognise that different SD specifications, which are often based on expert opinion where reported SD are missing, will yield sets of different, yet all valid, MRIOTs. This is an example of uncertainty proving to be useful within MRIO analysis.

Outcomes and Conclusions

With the recent release of several Global MRIO models and the observation that each model produces slightly different results, the development of a framework for understanding uncertainties in MRIO analysis is becoming an important area for future research. Calculations of uncertainty will provide an estimate of model robustness and reliability. Of particular interest would be an assessment of the suitability of MRIO model results for use in policy. The users of model results, who are not always the model developers, require an appreciation of the appropriateness of certain outcomes for use in policy applications. For example, recent studies have highlighted the increasing uncertainty associated with results at the sector level compared to national totals yet MRIO results have been and are increasingly being used to measure impacts at sector and product level. How to communicate uncertainty to users of MRIO outcomes is a further challenge for the research community.

Further Reading

Andrew, R., Peters, G. P., & Lennox, J. (2009). Approximation and Regional Aggregation in Multi-Regional Input-Output Analysis for National Carbon Footprint Accounting. *Economic Systems Research, 21* (3), 311-335.

Lenzen, M. (2011). Aggregation Versus Disaggregation in Input-Output Analysis of the Environment. *Economic Systems Research, 23* (1), 73-89.

Lenzen, M. (2000). Errors in Conventional and Input-Output—based Life—Cycle Inventories . *Journal of Industrial Ecology, 4* (4), 127-148.

Lenzen, M., Wood, R., & Wiedmann, T. (2010). Uncertainty analysis for Multi-Region Input-Output models - a case study of the UK's carbon footprint. *Economic Systems Research , 22*, 43-63.

Peters, G., Davis, S., & Andrew, R. (2012). A synthesis of carbon in international trade. *Biogeosciences Discussions , 9* (3), 3949-4023.

Wilting, H. (2012). Sensitivity and uncertainty analysis in MRIO modelling; some empirical results with regard to the Dutch carbon footprint. *Economic Systems Research , 24* (2), 141-171.

Part II: World MRIO Frameworks[1]

Lake Brewster…
Oil on canvas 127x 155cm
Dagmar Hoffmann

[1] Erik Dietzenbacher, Satoshi Inomata, Terrie Walmsley, and Arnold Tukker would like to acknowledge the assistance provided by the University of Sydney through a Round-11 International Program Development Fund (IPDF) grant enabling them to travel to Réunion in order to set up *Project Réunion* together with Manfred Lenzen, Bart Los and Thomas Wiedmann. The project aims to unite worldwide MRIO initiatives. They would also like to acknowledge the assistance provided by IDE-JETRO in Japan, which funded a second *Project Réunion* meeting.

Chapter 3: The Global Trade Analysis Project and the GTAP Database

Terrie L. Walmsley, Angel H. Aguiar, and Badri Narayanan

Introduction

The Global Trade Analysis Project (GTAP) is a global network of researchers and policy makers conducting quantitative analysis of international policy issues. The motivation and ultimate success of the Project stems from the fact that collaboration is essential for detailed analysis of the global economy. The importance of collaboration to improve the quality of policy analysis worldwide is most clearly seen in the development of the GTAP Data Base. The GTAP Data Base is the centerpiece of the Global Trade Analysis Project. It records the annual flows of goods and services for the entire world economy in the benchmark year(s). It consists of bilateral trade, transport, and protection matrices that link individual country/regional economic databases.

The production of the GTAP Data Base relies on the valuable contributions of many individuals and organizations throughout the world. Individuals contribute the best available input-output table for their country, while other experts contribute the macro, trade, protection and other data required. The Center for Global Trade Analysis (CGTA), the home of GTAP, then brings these contributions together into one useable, globally consistent, database. The result is a fully documented[1], publicly available and regularly updated global database. In 2012, the eighth version of the GTAP Data Base was released, covering 129 countries, 57 sectors, 5 factors and two base years.

---

[1] Angel Aguiar, Robert McDougall and Badri Narayanan G., Editors. *Global Trade, Assistance, and Production: The GTAP 8 Data Base*, Center for Global Trade Analysis, 2012 https://www.gtap.agecon.purdue.edu/databases/v8/v8_doco.asp

The GTAP Data Base is used in a suite of global comparative static and dynamic computable general equilibrium (CGE) models and underlies most contemporary economic analysis of global policy issues related to trade, energy and the environment. CGE models capture the economic relationships between sectors and/or regions through a set of simultaneous equations that describe the behaviour of the agents in the model (private households, governments, and firms), as well as the various markets for goods, services and factors of production (such as land, labour and capital). Underlying these equations are real data, which in combination with the model equations or theory, can be used to show how economies might react to changes in policy, technology or other external factors.[2] Collaboration in the development of the global database has allowed researchers to concentrate on the economic theory underlying these CGE models, which in turn has dramatically improved the quality of global economic analysis.

The purpose of this chapter is to introduce you to the GTAP Data Base and the Global Trade Analysis Project.

The GTAP Data Base and Satellite Datasets

The latest version of the GTAP Data Base, version 8, contains data on 129 countries[3], 57 sectors (Table 1) and 5 factors (land, skilled labor, unskilled labor, natural resources, and capital). For the first time in the history of the GTAP Data Base, there are also two base years: 2004 and 2007. In the following sections we first illustrate the structure of the GTAP Data Base, followed by a discussion of the individual components and the current set of satellite datasets, which accompany the GTAP Data Base.

---

[2] Comparative static CGE models model the reactions of an economy at a single point in time; while dynamic models attempt to trace the path of the economy/ies over time.

[3] The regions in the GTAP Data Base change with each release. A list of regions for version 8 with links to documentation is available at:

https://www.gtap.agecon.purdue.edu/databases/regions.asp?Version=8.211

Table 1: 57 Sectors in the GTAP 8 Data Base

Code	Description	Code	Description
PDR	Paddy rice	LUM	Wood products
WHT	Wheat	PPP	Paper products, publishing
GRO	Cereal grains	P_C	Petroleum, coal products
V_F	Vegetables, fruit, nuts	CRP	Chemical, rubber, plastic products
OSD	Oil seeds	NMM	Mineral products
C_B	Sugar cane, sugar beet	I_S	Ferrous metals
PFB	Plant-based fibers	NFM	Metals
OCR	Crops	FMP	Metal products
CTL	Bovine cattle, sheep and goats, horses	MVH	Motor vehicles and parts
OAP	Animal products	OTN	Transport equipment
RMK	Raw milk	ELE	Electronic equipment
WOL	Wool, silk-worm cocoons	OME	Machinery and equipment
FRS	Forestry	OMF	Manufactures
FSH	Fishing	ELY	Electricity
COA	Coal	GDT	Gas manufacture, distribution
OIL	Oil	WTR	Water
GAS	Gas	CNS	Construction
OMN	Minerals	TRD	Trade
CMT	Bovine meat products	OTP	Transport
OMT	Other Meat products	WTP	Water transport
VOL	Vegetable oils and fats	ATP	Air transport
MIL	Dairy products	CMN	Communication
PCR	Processed rice	OFI	Financial services
SGR	Sugar	ISR	Insurance
OFD	Food products	OBS	Business services
B_T	Beverages and tobacco products	ROS	Recreational and other services
TEX	Textiles	OSG	Public Administration, Defense, Education, Health
WAP	Wearing apparel	DWE	Dwellings
LEA	Leather products		

What does the GTAP Data Base Include?

Underlying the GTAP Data Base is a set of country/regional IO tables linked by international trade. Figure 1 illustrates the basic structure of the GTAP Data Base, excluding taxes. The arrays shown as solid rectangular boxes exist for each region. Since the dataset has multiple regions, we illustrate the other regions by the dashed arrays shown behind these solid boxes. Along the rows, each region's economy is summarized by the sales/uses of the 57 domestic and imported commodities (rows I and II respectively) and the 5 factors of production (row III); while inputs into production of the 57 commodities are shown in column I. The notation used matches the nomenclature used in the GTAP Data Base and model, and follows the format "V" for value; "D" or "I" for domestic or imports; "P",

"G", "F" or "X" for private, government, firm/intermediate or export demand respectively; and "M" for market prices (e.g., VIPM which means **V**alue of **I**mported purchases by **P**rivate households at **M**arket prices). In Figure 1 all values are at market prices, which are exclusive of taxes but inclusive of domestic margins. Certain values, such as sales of domestic (VDFM, VDPM, VDGM) and imported (VIFM, VIPM, VIGM) goods; and the sales of factors of production (VFM), are also shown in the GTAP Data Base inclusive of commodity or factor use taxes: signified by "A" for agent prices (e.g., VDFA which means **V**alue of **D**omestic purchases of **F**irms at **A**gents' prices). As is customary in I-O tables, total sales, exclusive of taxes, shown in the first row of Figure 1, should equal total costs, given by the sum of the first column, inclusive of taxes.

Uses also include sales of transport or margin commodities to the global transportation pool (VST), which are used to supply international transportations services for exporting goods from one country to another (VTWR in Figure 2). Three of the 57 GTAP commodities are also modes of transport (air, water and other transport) and correspond to the three margins in Figure 2. Figure 2 illustrates how the value of exports at market prices by destination (VXMD where "D" signifies destination) are linked to the value of imports at market prices by source country (VIMS where "S" signifies source) by export taxes/subsidies (XTAXD), transportation margins (VTWR) and import duties (MTAXD). Since GTAP was initially introduced as a trade model each of these wedges is separately identified by commodity (57), source (129) and destination (129), along with trade at f.o.b (freight on board) and c.i.f (cost inclusive of insurance and freight) prices.

Figure 1: Simplified view of the GTAP Data Base Structure (excluding commodity taxes)

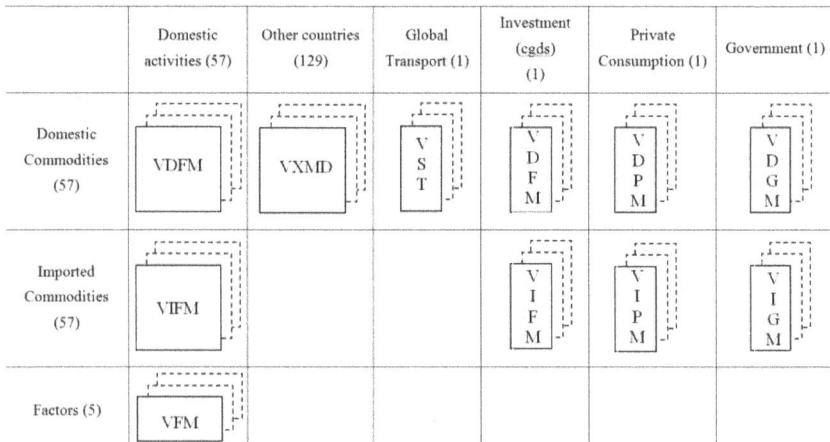

	Domestic activities (57)	Other countries (129)	Global Transport (1)	Investment (cgds) (1)	Private Consumption (1)	Government (1)
Domestic Commodities (57)	VDFM	VXMD	VST	VDFM	VDPM	VDGM
Imported Commodities (57)	VIFM			VIFM	VIPM	VIGM
Factors (5)	VFM					

N.B Commodity taxes are applied to all values except exports (VXMD and VST). Values inclusive of taxes would end in "A" for agent prices instead of "M" for market prices.

Figure 2: Link between Exports and Imports

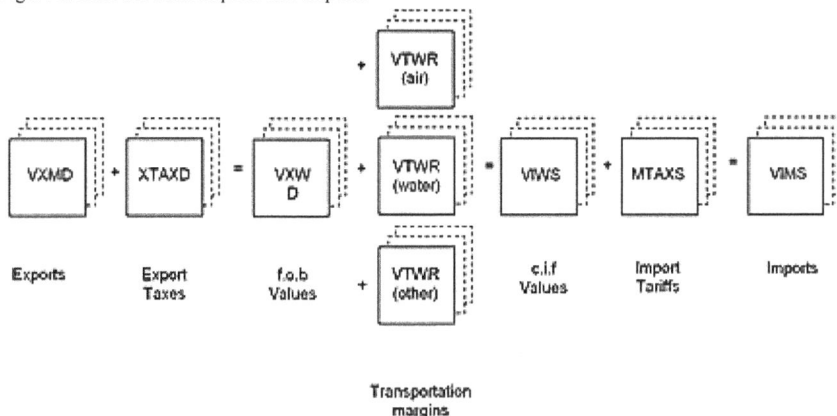

The value of imports by commodity, source and destination at market prices (VIMS) are equal to the value of imports by commodity purchased by firms (VIFM), government (VIGM), investment (VIFM) and private consumption (VIPM), also at market prices. Notice that the GTAP Data Base does not separate imports by source and agent, although versions of GTAP in this MRIO structure are available (more details below).

In addition, the GTAP Data Base also includes income taxes, savings, capital stocks, depreciation, and population data for each country/region.

What does the GTAP Data Base look like?

When you receive the GTAP Data Base you obtain the GTAPAgg program, a user-friendly aggregation program (Figure 3) which is used to create your own aggregation. The data are in Millions of US dollars and are contained in a header array file.[4] Summaries tables and tax rates are also provided, as well as the data set out in a useful Social Accounting Matrix (SAM) structure.[5]

---

[4] GEMPACK software suite: Harrison, J. and K. R. Pearson. 2007. *GEMPACK User Documentation Release 8.0.* Centre of Policy Studies and Impact Project Monash University, Melbourne, Australia.
[5] McDonald S. and K. Theirfelder, "Deriving a Global Social Accounting Matrix from GTAP Data Base" *GTAP Technical Paper,* 22, Global Trade Analysis Project, 2004. https://www.gtap.agecon.purdue.edu/resources/res_display.asp?RecordID=1645

Figure 3: GTAPAgg

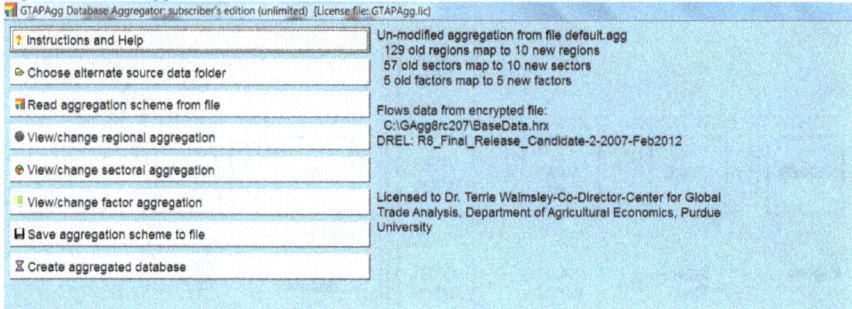

Where do the Data in the GTAP Data Base come from?

All of the datasets used in the GTAP Data Base are contributed by network members. Contributors provide their country's IO data in a particular format[6]: tables must balance and should include at least 30 sectors. Since IO tables are usually only produced by a country's statistical office every 5 to 10 years, we do not require that contributed IO tables match the base year of the GTAP Data Base, instead we update the table to the relevant year ourselves using various macro datasets.

The quality of IO Tables can vary considerably across countries. While contributors are expected to provide as much data as possible, over the years the Center has had to collect a number of additional data to reinforce the quality of the IO tables. For instance, many IO tables do not have the level of agricultural detail required and hence additional agricultural IO data are collected from the FAO and processed by Everett Peterson (Professor, Virginia Tech University, USA) to disaggregate agricultural where required. Energy has also been identified as an area where additional processing is required to ensure consistency with global data sources (e.g., Energy Statistics from the OECD's International Energy Agency). Other areas where the Center collects additional data to reinforce the IO tables include income taxes (collected from the IMF) and land usage by sector.

Macroeconomic data are obtained from the World Bank and are used to ensure that the country IO data match the macro data for the base year. Trade Data is taken from the UN COMTRADE database, reconciled by Mark Gehlhar (Economist, US Department of Interior), with some adjustments made for re-exports. Services trade data are also compiled by Nico van Leeuwen and Arjan Lejour (Data Analyst and Project Leader, Netherlands Bureau for Economic Policy Analysis) using data collected from the OECD and IMF. Protection data is contributed by CEPII and the ITC-Geneva, export subsidies by David Laborde (Senior Research Fellow, International Food Policy Research Institute) and domestic support by the OECD and Hans Jensen (Researcher, Institute of Food and Resource Economics, University of Copenhagen).

---

[6] Huff, K., R. McDougall and T. Walmsley. Contributing Input-Output Tables to the GTAP Data Base *GTAP Technical Paper,* 12, Global Trade Analysis Project, 2000. Resources for contributors:
https://www.gtap.agecon.purdue.edu/databases/contribute/default.asp.

Ensuring Consistency Amongst the Data

The GTAP Data Base is a globally consistent database. This means that all the data are internally consistent, despite coming from different sources. Moreover the GTAP Data Base must satisfy certain adding up constraints that one would expect to see within an economy and across the world as a whole.

When the data are first collected they are not globally consistent. Global exports do not match global imports. Even on a bilateral basis, exports to Tanzania reported by the USA, for instance, are unlikely to match imports from the USA reported by Tanzania. Even after transportation costs and taxes are taken into account there are still differences between the values because of data collection issues. Moreover these trade data should also match the imports reported in each individual country's I-O table, although they rarely do.

While such data issues are considered normal, general equilibrium models require that the data match and there are no leakages (i.e., for example supply must always equal demand for all commodities). In order to reconcile these different data sources, decisions need to be made about which data to believe and which data should be adjusted to ensure this consistency. These decisions are based on our beliefs about the relative reliability of the different data sources and expert advice.

In general we require that each contributor of a dataset ensure that their dataset is internally consistent before it is sent to us: for example, in the contributed trade dataset, exports must equal imports. This reflects the fact that the contributors know their datasets better than we do and hence are in a better position to make decisions about the quality of inconsistent data. For instance, Mark Gehlhar (US Department of Interior) produces a globally consistent goods trade database using reliability indexes.[7]

Our task is then to ensure consistency amongst the different datasets contributed. To do this we rank the reliability of the datasets: the goods trade data is assumed to be the most reliable dataset, followed by the macro data, energy data etc. The FIT process is used to 'fit' the data using entropy theoretic methods.[8] The IO tables therefore undergo changes during this FIT process to reflect any differences between the table and the additional macroeconomic data. The extent of the changes depends on the age and quality of the underlying IO table – good tables generally match the macro data more closely and therefore changes are minor.

What other Data are available with the GTAP Data Base?

The GTAP Data Base also includes a number of other datasets needed for modeling purposes, as well as a number of GTAP satellite datasets, developed as a result of research projects designed to extend the GTAP Data Base and model.

---

[7] Gehlhar, M., "Reconciling Bilateral Trade Data for Use in GTAP" *GTAP Technical Paper*, 10, Global Trade Analysis Project, 1996.
https://www.gtap.agecon.purdue.edu/resources/res_display.asp?RecordID=313
[8] James, M. and R. McDougall, "FIT: An input-output data update facility for SALTER", *SALTER Working Paper,* 17, Industry Commission, Canberra, 1993.

1. Elasticities, required by the GTAP model, defining the extent to which demand responds to changes in prices and/or incomes. These are taken from econometric estimates obtained from the literature.
2. Time series trade data showing the bilateral goods trade valued at f.o.b (VXWD) between 1992 and 2009.
3. Energy volumes and CO2 emissions by use for all energy commodities (coal, oil, gas, petroleum and coal products, electricity and gas distribution). Energy volumes and CO2 emissions are in million tons of equivalents.
4. The Dynamic GTAP (GDyn) Data Base contains additional data on foreign income flows, rates of return, as well as estimates of the speed at which rates of return converge globally over time, required for the Dynamic GTAP model.[9]
5. The Migration (GMig2) Data Base contains additional data on numbers of migrant workers by origin, wages and remittances required for the global migration model.[10]
6. The land use database contains disaggregated land data by use. Land is disaggregated into 26 Agro ecological zones. [11] The 26 types of land/zones differ depending on availability of water and soil quality, and hence the productivity and suitability of the land for the production of agricultural crops and/or livestock.

Tools for Manipulating the GTAP Data Base

New tools are being developed all the time by GTAP network members to expand and manipulate the GTAP Data Base. Below is a list of some of the tools available:

-   Elasticities can be altered directly by changing their value in the parameters file using ViewHAR.
-   The 57 GTAP Commodities can be further disaggregated using the SplitCom program.[12]
-   Tax rates in the GTAP Data Base can be altered using a special simulation called ALTERTAX.[13]

---

[9] Ianchovichina, E.; and Walmsley, T. L., (eds): *Dynamic Modeling and Applications in Global Economic Analysis,* Cambridge University Press, 2012.

[10] Walmsley, T., A. Ahmed and C. Parsons, "The GMig2 Data Base: A Data Base of Bilateral Labor Migration, Wages and Remittances", *GTAP Research Memorandum,* 16, 2005. https://www.gtap.agecon.purdue.edu/resources/res_display.asp?RecordID=1880

[11] Avetisyan, M., U. Baldos, T. Hertel, "Development of the GTAP Version 7 Land Use Data Base" *GTAP Research Memorandum,* 19, Global Trade Analysis Project, 2011. https://www.gtap.agecon.purdue.edu/resources/res_display.asp?RecordID=3426

[12] Horridge, M., SplitCom: Programs to disaggregate a GTAP sector Centre of Policy Studies, Monash University, 2008.
https://www.gtap.agecon.purdue.edu/resources/splitcom.asp

[13] Malcom, G., Adjusting Tax Rates in the GTAP Data Base, *GTAP Technical Paper,* 12, 1998.
https://www.gtap.agecon.purdue.edu/access_staff/resources/res_display.asp?RecordID=31 5

   -   Detailed trade and tariff data can be viewed and manipulated in the TASTE program.[14]

GTAP MRIO

The GTAP Data Base is not in the MRIO format outlined in chapter 1, because it does not separate imports of each commodity by agent, source and destination. Instead we account for imports of each commodity purchased by agent in each destination (VIFM, VIGM, VIPM); and for total imports of each commodity by source and destination (VIMS). A number of people in the GTAP network[15] have taken the GTAP Data Base and used shares from the commodity, destination, and source specific import data (VIMS in Figure 2) to further split imports by agent (VIPM, VIGM, VIFM in Figure 1) across sources: thereby producing a basic MRIO. For example, this work splits USA imports of Coal from China into imports by firms, government and private households using the average share of US coal imports coming from China. Recently, Marinos Tsigas and Zhi Wang (US International Trade Commission), and Mark Gehlhar (US Department of Interior)[16] used additional information from the trade data to construct a GTAP MRIO that more accurately depicts the differences between agents' sourcing of imports.

Uses

The GTAP Data Base is used extensively for computable general equilibrium (CGE) modeling, and by Glen Peters (Senior research fellow, CICERO, Norway) and his colleagues for multi-region input-output modeling. CGE models are descended from these input-output models, which were first pioneered by Wassily Leontief. Underlying both CGE and I-O models are the inter-industry relations within an economy, first depicted by Leontief in an input-output matrix. The main difference between the models is that CGE models assign a larger role to prices and resource constraints.

---

[14] Horridge, M. and D. Laborde, "TASTE: a program to adapt detailed trade and tariff data to GTAP-related purposes" *Presented at the 11th Annual Conference on Global Economic Analysis*, Helsinki, Finland, 2008.
https://www.gtap.agecon.purdue.edu/resources/taste/taste.asp
[15] Peters, G., R. Andrew, and J. Lennox. "Constructing an Environmentally extended Multi-Regional Input–Output Table Using the GTAP Data Base." *Economic Systems Research, 2011, Vol. 23(2), June, pp. 131–152*, 2011; Rutherford, T., J. Carbone, and C. Böhringer. "Using Embodied Carbon To Control Carbon Leakage." *Presented at the 14th Annual Conference on Global Economic Analysis, Venice, Italy*, 2011; and Reimer, Jeff. "The Domestic Content of Imports and the Foreign Content of Exports." *International Review of Economics and Finance*, 2010.
[16] Tsigas, M., Z. Wang and M. Gehlhar, How a Global Inter-Country Input-Output Table with Processing Trade Account can be constructed from GTAP Database, to be presented at *2012 Conference on Global Economic Analysis*, Geneva, 2012.
https://www.gtap.agecon.purdue.edu/access_staff/resources/res_display.asp?RecordID=37
84

The GTAP Data Base underlies most of the global CGE models, including the GTAP model.[17] The GTAP model is a relatively standard multi-region, CGE model that is easy to use and modify. A number of model extensions have been developed to include alternative forms of competition amongst firms [18], technological spillovers across countries[19], CO_2 emissions[20], land use[21], poverty[22], migration[23] and dynamics[ix]. There are currently over 2,000 applications on the GTAP website, representing just a small fraction of the total number of applications undertaken using the GTAP Data Base.

The GTAP MRIO developed by Glen Peters has been used to trace value added in global production chains[24], in the OPEN:EU project to assess the environmental footprints of products[25], and in the Social HotSpots Database initiative to quantify the social impacts of a product over its life cycle.[26]

---

[17] Hertel, T., ed., *Global Trade Analysis: Modeling and Applications*, Cambridge University Press, 1997.

[18] Francois, J., "Scale Economies and Imperfect Competition in the GTAP Model", *GTAP Technical Paper*, 14, Global Trade Analysis Project, 1998.
https://www.gtap.agecon.purdue.edu/resources/res_display.asp?RecordID=317

[19] Van Meijl, H., and Van Tongeren, F., Endogenous International Technology Spillovers and Biased Technical Change in the GTAP Model", *GTAP Technical Paper*, 15, Global Trade Analysis Project, 1999.
https://www.gtap.agecon.purdue.edu/resources/res_display.asp?RecordID=318

[20] Burniaux, J., and T. Truong, "GTAP-E: An Energy-Environmental Version of the GTAP Model", *GTAP Technical Paper*, 16, Global Trade Analysis Project, 2002.
https://www.gtap.agecon.purdue.edu/resources/res_display.asp?RecordID=923

[21] Keeney, R., and T. Hertel, "GTAP-AGR: A Framework for Assessing the Implications of Multilateral Changes in Agricultural Policies", *GTAP Technical Paper*, 24, Global Trade Analysis Project, 2005.
https://www.gtap.agecon.purdue.edu/resources/res_display.asp?RecordID=1869

[22] Hertel, T., M. Verma, M. Ivanic and A. R. Rios, "GTAP-POV: A Framework for Assessing the National Poverty Impacts of Global Economic and Environmental Policies", *GTAP Technical Paper*, 31, 2011,
 https://www.gtap.agecon.purdue.edu/resources/res_display.asp?RecordID=3731

[23] Walmsley, T. L., A. Winters, and S. A., Ahmed, " Measuring the Impact of the Movement of Labor Using a Model of Bilateral Migration Flows", *GTAP Technical Paper*, 28, Global Trade Analysis Project, 2007.
https://www.gtap.agecon.purdue.edu/resources/res_display.asp?RecordID=2529

[24] Koopman, R., W. Powers, Z. Wang, S. Wei, "Give Credit where credit is due: Tracing value added in Global production chains, *NBER Working Paper*, 2010.

[25] Hertwich E. and G. Peters. "Multiregional Input-Output Database" *OPEN:EU Technical Document*, Trondheim, 2010.
 http://www.oneplaneteconomynetwork.org/resources/programme-
documents/WP1_MRIO_Technical_Document.pdf

[26] Benoit C., G. Norris, D. Aulisio, S. Rogers, J. Reed, and S. Overaker, "Risk and Opportunity Table Development" New Earth, 2010.
http://socialhotspot.org/userfiles/SHDB_Risk_And_Opportunity_Tables_-
_Final_Fall_2010.pdf

A Short History of Development

The GTAP 8 Data Base builds heavily on earlier work at Purdue, as well as research and database development efforts at a number of national and international agencies. Indeed, the earliest versions were developed as part of the SALTER Project, undertaken by the Australian Industry Commission in the 1980s and early 90s. Table 2 below shows the gradual expansion of regional and sectoral coverage, along with some of the other major improvements in the database.

Table 2: Timeline of GTAP Data Base Releases

Release	Released	Regions	Sectors	Base Year	Other improvements
1	1993	15	37	1990	
2	1994	24	37	1992	
3	1996	30	37	1992	
4	1998	45	50 (more agriculture added)	1995	Physical energy flows, International Energy Agency. Breakout of skilled and unskilled labor.
5	2001	66	57 (more services added)	1997	Agriculture Market Analysis Database (AMAD), ERS/USDA; and World Integrated Trade Software (WITS), World Bank and UNCTAD for tariffs.
6	2005	87	57	2001	MAcMap data base for tariffs and import protection, CEPII and ITC-Geneva.
7	2008	113	57	2004	Bilateral Services trade data, CPB.
8	2012	129	57	2004 2007	Inclusion of CO2 emissions and land use.

Further reading

Hertel, T. (2012). Global Applied General Equilibrium Analysis using the GTAP Framework. In P. B. Dixon, & D. Jorgenson, *Handbook of Computable General Equilibrium Modeling,* part of *Handbook of Economics Series* (Vol. 1B). Elsevier.

Powell, A. A. (2007). Why, How and When did GTAP Happen? What has it Achieved? Where is it Heading? GTAP Working Paper 38, Global Trade Analysis PRoject. Purdue University. West Lafayette: Purdue University, Indiana.

Additional Resources

https://www.gtap.agecon.purdue.edu/

Acknowledgements

The authors would like to acknowledge Thomas Hertel for his comments.

Chapter 4: Transnational Interregional Input-Output Tables: An Alternative Approach to MRIO?

Satoshi Inomata and Bo Meng

Introduction

The Institute of Developing Economies (IDE-JETRO) has been constructing international input-output tables for more than 30 years. These datasets are well-recognised as a powerful analytical tool for the study of industrial structure and production networks across countries, and thus have earned the enthusiastic attention of users world-wide.

In the last few years, however, some prominent academic institutions and international organisations have simultaneously launched new projects for constructing MRIO tables, such as the World Input-Output Database (WIOD) of the European Commission, the OECD international input-output database by the Organisation for Economic Co-operation and Development, or the Eora database by the University of Sydney, as introduced in other sections of this book.

While this clearly reflects an increasing demand for MRIO databases, it also implies that each project should endeavour to differentiate its own product (i.e. MRIO data) from others in order to meet the multiple needs of users.

One of the possible ways for a project to promote its competitiveness is to broaden the country coverage, so that it can appeal to a wider range of users, say, from trade specialists to environmental economists. This challenge however entails significant monetary and non-monetary cost, as it is known that the more countries the table covers, the more difficult it will be to harmonise constituent national tables into a common format.

Another strategy is to specialise in a limited number of countries, yet enriching the tables with specific regional information. IDE-JETRO's new project of constructing transnational interregional input-output tables (TIIO) pursues this direction. The basic idea of the scheme is to link interregional input-output tables of several countries into a single matrix using the import/export data of the regional customs of individual countries.

IDE-JETRO has already constructed and released in 2007 the 2000 Transnational Interregional Input-Output Table between China and Japan. The table enabled us to study economic linkages across borders on a region-to-region basis, say, between Huanan region in China and Kyushu region in Japan1. Thanks to favourable feedback from many users, a new project of constructing the 2005 Transnational Interregional Input-Output Table for China, Japan and Korea (Figure 1) was launched, using the basic framework of the newly released 2005 Asian International Input-Output Table. It is expected that the data will serve to draw a detailed mapping of cross-national production networks in East Asia, in particular of the dynamics of regional industrial structures after China's accession to the WTO in 2001.

The project will be conducted in a close collaboration among institutions in charge of constructing their own national tables: i.e., the State Information Center of China, the Bank of Korea and the Ministry of Economy, Trade and Industry of Japan.

The Tool and its Data Specification

Data Layout

In Figure 1, there are three categories of country groups in the table. Core countries refer to the countries with regional disaggregation; that is, the countries whose interregional input-output tables are linked in the TIIO framework. For the case of TIIO2005, they are mainland China, Japan, and the Republic of Korea. Endogenous countries/regions are countries/regions whose national input-output tables are individually integrated into the system, so that the inter-country Leontief inverse matrix can also be derived with respect to their intermediate transactions. These are Taiwan, ASEAN 5 (Indonesia, Malaysia, the Philippines, Singapore and Thailand) as one region, and the USA. Any other countries not included in the above are aggregated into the Rest of the World, presented as a matrix for the input (import) side and as a vector for the output (export) side.

---

[1]Currently the Asian International Input-Output Table covers only ten countries, but its classification is as detailed as 76 products, which well surpasses the specification of other MRIO tables. For the year 2005, IDE-JETRO also constructed an international input-output table for BRICs economies, covering Brazil, Russia, India, China, the USA, Japan and EU27 (as a single region).

Figure 1: Image of 2005 Transnational Interregional Inut-Output Table for China, Japan, and Korea

			Core countries									Endogenous countries / regions			Core countries									Endogenous countries / regions			ROW	
			China-region 1	..	China-region 7	Japan-region 1	..	Japan-region 9	Korea-region 1	..	Korea-region 4	Taiwan	ASEAN 5	USA	China-region 1	..	China-region 7	Japan-region 1	..	Japan-region 9	Korea-region 1	..	Korea-region 4	Taiwan	ASEAN 5	United States	Export to the Rest of the World	Total Output
		Code: regional (4 digits)	AC01001 : AC01015		AC07001 : AC07015	AJ01001 : AJ01015		AJ09001 : AJ09015	AK01001 : AK01015		AK04001 : AK04015	AN00001 : AN00015	AQ00001 : AQ00015	AU00001 : AU00015	FC01001 : FC01004		FC07001 : FC07004	FJ01001 : FJ01004		FJ09001 : FJ09004	FK01001 : FK01004		FK04001 : FK04004	FN00001 : FN00004	FQ00001 : FQ00004	FU00001 : FU00004	LW00	XX00
Core countries	China-region 1	AC01001 : AC01015	A^{C01C01}	..	A^{C01C07}	A^{C01J01}	..	A^{C01J09}	A^{C01K01}	..	A^{C01K04}	A^{C01N00}	A^{C01Q00}	A^{C01U00}	F^{C01C01}	..	F^{C01C07}	F^{C01J01}	..	F^{C01J09}	F^{C01K01}	..	F^{C01K04}	F^{C01N00}	F^{C01Q00}	F^{C01U00}	L^{C01W00}	X^{C01}
	:		:	:	:	:	:	:	:	:	:	:	:	:	:	:	:	:	:	:	:	:	:	:	:	:	:	:
	China-region 7	AC07001 : AC07015	A^{C07C01}	..	A^{C07C07}	A^{C07J01}	..	A^{C07J09}	A^{C07K01}	..	A^{C07K04}	A^{C07N00}	A^{C07Q00}	A^{C07U00}	F^{C07C01}	..	F^{C07C07}	F^{C07J01}	..	F^{C07J09}	F^{C07K01}	..	F^{C07K04}	F^{C07N00}	F^{C07Q00}	F^{C07U00}	L^{C07W00}	X^{C07}
	Japan-region 1	AJ01001 : AJ01015	A^{J01C01}	..	A^{J01C07}	A^{J01J01}	..	A^{J01J09}	A^{J01K01}	..	A^{J01K04}	A^{J01N00}	A^{J01Q00}	A^{J01U00}	F^{J01C01}	..	F^{J01C07}	F^{J01J01}	..	F^{J01J09}	F^{J01K01}	..	F^{J01K04}	F^{J01N00}	F^{J01Q00}	F^{J01U00}	L^{J01W00}	X^{J01}
	:		:	:	:	:	:	:	:	:	:	:	:	:	:	:	:	:	:	:	:	:	:	:	:	:	:	:
	Japan-region 9	AJ09001 : AJ09015	A^{J09C01}	..	A^{J09C07}	A^{J09J01}	..	A^{J09J09}	A^{J09K01}	..	A^{J09K04}	A^{J09N00}	A^{J09Q00}	A^{J09U00}	F^{J09C01}	..	F^{J09C07}	F^{J09J01}	..	F^{J09J09}	F^{J09K01}	..	F^{J09K04}	F^{J09N00}	F^{J09Q00}	F^{J09U00}	L^{J09W00}	X^{J09}
	Korea-region 1	AK01001 : AK01015	A^{K01C01}	..	A^{K01C07}	A^{K01J01}	..	A^{K01J09}	A^{K01K01}	..	A^{K01K04}	A^{K01N00}	A^{K01Q00}	A^{K01U00}	F^{K01C01}	..	F^{K01C07}	F^{K01J01}	..	F^{K01J09}	F^{K01K01}	..	F^{K01K04}	F^{K01N00}	F^{K01Q00}	F^{K01U00}	L^{K01W00}	X^{K01}
	:		:	:	:	:	:	:	:	:	:	:	:	:	:	:	:	:	:	:	:	:	:	:	:	:	:	:
	Korea-region 4	AK04001 : AK04015	A^{K04C01}	..	A^{K04C07}	A^{K04J01}	..	A^{K04J09}	A^{K04K01}	..	A^{K04K04}	A^{K04N00}	A^{K04Q00}	A^{K04U00}	F^{K04C01}	..	F^{K04C07}	F^{K04J01}	..	F^{K04J09}	F^{K04K01}	..	F^{K04K04}	F^{K04N00}	F^{K04Q00}	F^{K04U00}	L^{K04W00}	X^{K04}
Endogenous countries / regions	Taiwan	AN00001 : AN00015	A^{N00C01}	..	A^{N00C07}	A^{N00J01}	..	A^{N00J09}	A^{N00K01}	..	A^{N00K04}	A^{N00N00}	A^{N00Q00}	A^{N00U00}	F^{N00C01}	..	F^{N00C07}	F^{N00J01}	..	F^{N00J09}	F^{N00K01}	..	F^{N00K04}	F^{N00N00}	F^{N00Q00}	F^{N00U00}	L^{N00W00}	X^{N00}
	ASEAN 5	AQ00001 : AQ00015	A^{Q00C01}	..	A^{Q00C07}	A^{Q00J01}	..	A^{Q00J09}	A^{Q00K01}	..	A^{Q00K04}	A^{Q00N00}	A^{Q00Q00}	A^{Q00U00}	F^{Q00C01}	..	F^{Q00C07}	F^{Q00J01}	..	F^{Q00J09}	F^{Q00K01}	..	F^{Q00K04}	F^{Q00N00}	F^{Q00Q00}	F^{Q00U00}	L^{Q00W00}	X^{Q00}
	USA	AU00001 : AU00015	A^{U00C01}	..	A^{U00C07}	A^{U00J01}	..	A^{U00J09}	A^{U00K01}	..	A^{U00K04}	A^{U00N00}	A^{U00Q00}	A^{U00U00}	F^{U00C01}	..	F^{U00C07}	F^{U00J01}	..	F^{U00J09}	F^{U00K01}	..	F^{U00K04}	F^{U00N00}	F^{U00Q00}	F^{U00U00}	L^{U00W00}	X^{U00}
International freight & insurance		BF00	BA^{C01}	..	BA^{C07}	BA^{J01}	..	BA^{J09}	BA^{K01}	..	BA^{K04}	BA^{N00}	BA^{Q00}	BA^{U00}	BF^{C01}	..	BF^{C07}	BF^{J01}	..	BF^{J09}	BF^{K01}	..	BF^{K04}	BF^{N00}	BF^{Q00}	BF^{U00}		
ROW	Import from the Rest of the World	CW00001 : CW00015	A^{W00C01}	..	A^{W00C07}	A^{W00J01}	..	A^{W00J09}	A^{W00K01}	..	A^{W00K04}	A^{W00N00}	A^{W00Q00}	A^{W00U00}	F^{W00C01}	..	F^{W00C07}	F^{W00J01}	..	F^{W00J09}	F^{W00K01}	..	F^{W00K04}	F^{W00N00}	F^{W00Q00}	F^{W00U00}		
Duties & Import commodity taxes		DT00	DA^{C01}	..	DA^{C07}	DA^{J01}	..	DA^{J09}	DA^{K01}	..	DA^{K04}	DA^{N00}	DA^{Q00}	DA^{U00}	DF^{C01}	..	DF^{C07}	DF^{J01}	..	DF^{J09}	DF^{K01}	..	DF^{K04}	DF^{N00}	DF^{Q00}	DF^{U00}		
Value Added		VV001 : VV004	V^{C01}	..	V^{C07}	V^{J01}	..	V^{J09}	V^{K01}	..	V^{K04}	V^{N00}	V^{Q00}	V^{U00}														
Total Input		XX00	X^{C01}	..	X^{C07}	X^{J01}	..	X^{J09}	X^{K01}	..	X^{K04}	X^{N00}	X^{Q00}	X^{U00}														

* Each cell of A** and F** represents a matrix of 15 x 15 and of 15 x 4 dimension, respectively.

Reading the table in column-wise direction, each cell shows input compositions of industries of respective (sub-)regions. For example, AC01C01 shows the input compositions of China-region01's industries vis-à-vis their own goods and services, i.e. intra-regional transaction of China-region01. AC07C01 in contrast shows the input compositions of China-region01 for the products produced by its domestic neighbour China-region07, i.e. inter-regional transaction within China. Further down, AJ01C01 presents China-region01's intermediate demand for imported products from Japan-region01. Other lower cells in the column allow the same interpretation for China's imports from other foreign regions.

Turning to the right-half of the table, F***C01 shows the flow of products into final demand sectors of China-region01. FC01C01 and FJ01C01, for example, map the compositions of China-region01's final demand, for goods and services produced within the region and for those imported from Japan-region01, respectively. The rest of the column is read in the same manner as for the 1st column of the table.

BA and DA give international freight & insurance margins and taxes on import transaction, while LW, VV and XX are export, value-added and total input/output, as seen in the conventional national I-O table.

Regional Coverage

Coverage of sub-regions within a country is shown in Figure 2. Note that Okinawa in Japan is presented as a separate region in TIIO2005, while it was merged with Kyushu in the previous TIIO2000. This change reflects a growing importance of Okinawa region as a logistic hub for international distribution networks in East Asia.

Industrial Sector Classification

Table 1 shows concordance tables between TIIO and the Asian International Input-Output Table (AIIO) with 76 sectors. There are some notable changes in industrial sector classification from TIIO2000 to TIIO2005. Firstly, the number of sectors has been increased from 10 to 15. Sector codes in TIIO2005 are appended with sub-codes (a, b, c …etc.) so that they can be cross-referred to those in TIIO2000. Also, product coverage (vis-à-vis AIIO sectors) is redefined in TIIO2005 to better elucidate the structure of vertical production chains, while the classification of TIIO2000 was designed in accordance with end-use categories.

Figure 2: Regional coverage

Sub-regions in China
C01 Dongbei: Liaoning(6), Jilin(7), Heilongjiang(8)
C02 Huabei: Beijing(1), Tianjin(2), Hebei(3), Shandong(15)
C03 Huadong: Shanghai(9), Jiangsu(10), Zhejiang(11)
C04 Huanan: Fujian(13), Guangdong(19), Hainan(21)
C05 Huazhong: Shanxi(4), Anhui(12), Jiangxi(14), Henan(16), Hubei(17), Hunan(18)
C06 Xibei: Inner Mongolia(5), Shaanxi,(27), Gansu(28), Qinghai(29), Ningxia(30), Xinjiang(31)
C07 Xinan: Guangxi(20), Chongqing(22), Sichuan(23), Guizhou(24), Yunnan(25), Tibet(26)

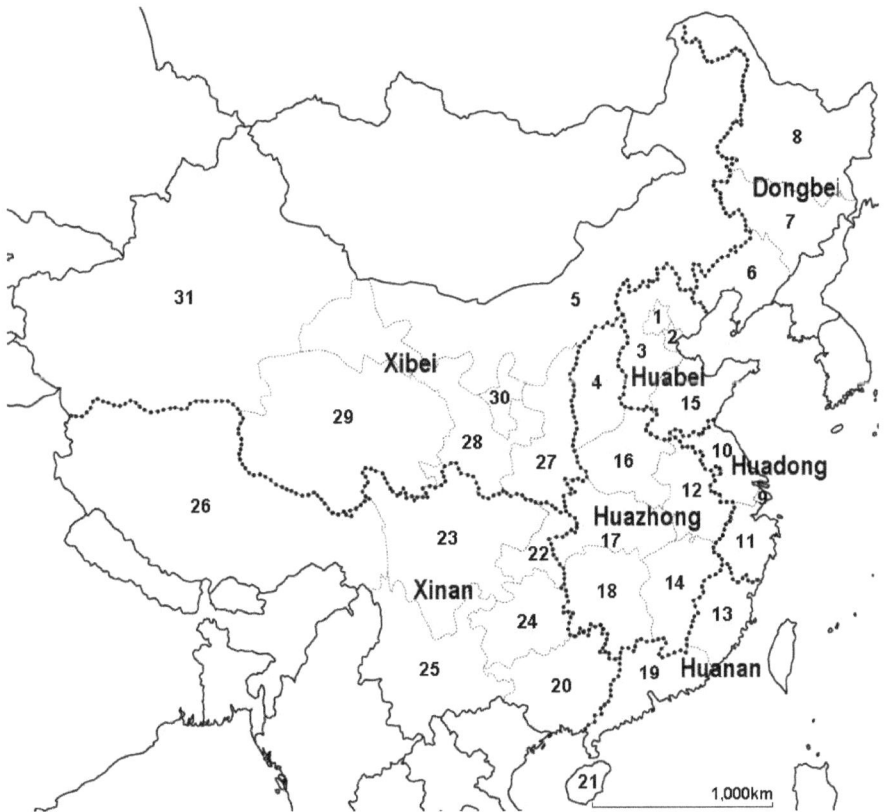

Sub-regions in Korea

K01 Sudokwon: Seoul(1), Incheon(2), Gyeonggi-do(3)
K02 Jungbukwon: Daejeon(4), Gangwon-do(5),
    Chungcheongbuk-do(6), Chungcheongnam-do(7)
K03 Yeongnamkwon: Daegu(8), Ulsan(9), Busan(10),
    Gyeongsangbuk-do(11), Gyeongsangnam-do(12)
K04 Honamkwon: Gwangju(13), Jeollabuk-do(14),
    Jeollanam-do(15), Jeju-do(16)

Sub-regions in Japan

J01 Hokkaido: Hokkaido(1)
J02 Tohoku: Aomori(2), Iwate(3), Miyagi(4), Akita(5), Yamagata(6),
    Fukushima(7)
J03 Kanto: Ibaraki(8), Tochigi(9), Gunma(10), Saitama(11), Chiba(12),
    Tokyo(13), Kanagawa(14), Niigata(15), Yamanashi(19),
    Nagano(20), Shizuoka(22)
J04 Chubu: Toyama(16), Ishikawa(17), Gifu(21), Aichi(23), Mie(24)
J05 Kinki: Fukui(18), Shiga(25), Kyoto(26), Osaka(27), Hyogo(28),
    Nara(29), Wakayama(30)
J06 Chugoku: Tottori(31), Shimane(32), Okayama(33), Hiroshima(34),
    Yamaguchi(35)
J07 Shikoku: Tokushima(36), Kagawa(37), Ehime(38), Kochi(39)
J08 Kyushu: Fukuoka(40), Saga(41), Nagasaki(42), Kumamoto(43),
    Oita(44), Miyazaki(45), Kagoshima(46)
J09 Okinawa: Okinawa(47)
    *J08 and J09 were merged as one region in TIIO2000

Sources: State of Information Center of China, Bank of Korea, IDE-JETRO

76 Sector Classification of the 2000 AIO Table

15 Sector Classification		76 Sector Classification of the 2000 AIO Table	
Code	Description	Code	Description
001	Agriculture, livestock, forestry	001	Paddy
		002	Other grain
		003	Food crops
		004	Non-food crops
		005	Livestock and poultry
		006	Forestry
		007	Fishery
002	Mining and quarrying	008	Crude petroleum and natural gas
		009	Iron ore
		010	Other metallic ore
		011	Non-metallic ore and quarrying
003	Household consumption pro	012	Milled grain and flour
		013	Fish products
		014	Slaughtering, meat products and dairy products
		015	Other food products
		016	Beverage
		017	Tobacco
		018	Spinning
		019	Weaving and dyeing
		020	Knitting
		021	Wearing apparel
		022	Other made-up textile products
		023	Leather and leather products
		024	Timber
		025	Wooden furniture
		026	Other wooden products
		060	Other manufacturing products
		027	Pulp and paper
		028	Printing and publishing
		029	Synthetic resins and fiber
		030	Basic industrial chemicals
		031	Chemical fertilizers and pesticides
		032	Drugs and medicine
		033	Other chemical products
004	Basic industrial materials	034	Refined petroleum and its products
		035	Plastic products
		036	Tires and tubes
		037	Other rubber products
		038	Cement and cement products
		039	Glass and glass products
		040	Other non-metallic mineral products

15 Sector Classification		76 Sector Classification of the 2000 AIO Table	
Code	Description	Code	Description
		041	Iron and steel
		042	Non-ferrous metal
		043	Metal products
		044	Boilers, engines and turbines
		045	General machinery
		046	Metal working machinery
		047	Specialized machinery
		048	Heavy electrical equipment
005	Processing and assembling	049	Television sets, radios, audios and communication equipment
		050	Electronic computing equipment
		051	Semiconductors and integrated circuits
		052	Other electronics and electronic products
		053	Household electrical equipment
		054	Lighting fixtures, batteries, wiring and others
		055	Motor vehicles
		056	Motor cycles
		057	Shipbuilding
		058	Other transport equipment
		059	Precision machines
006	Electricity, gas and water sup	061	Electricity and gas
		062	Water supply
007	Construction	063	Building construction
		064	Other construction
008	Trade	065	Wholesale and retail trade
009	Transportation	066	Transportation
010	Services	067	Telephone and telecommunication
		068	Finance and insurance
		069	Real estate
		070	Education and research
		071	Medical and health service
		072	Restaurant
		073	Hotel
		074	Other services
		075	Public administration
		076	Unclassified

Value Added and Final Demand Items

Value Added		Final Demand	
001	Wages and salary	001	Private consumption
002	Operating surplus	002	Government consumption
003	Depreciation of fixed capital	003	Gross fixed capital formation
004	Indirect taxes less subsidies	004	Changes in stocks

TIIO2005 (Differences from TIIO2000 are shown in bold.)

15 Sector Classification		76 Sector Classification of the 2000 AIO Table	
Code	**Description**	**Code**	**Description**
001	Agriculture, livestock, forestry, and fishery	001	Paddy
		002	Other grain
		003	Food crops
		004	Non-food crops
		005	Livestock and poultry
		006	Forestry
		007	Fishery
002	Mining and quarrying	008	Crude petroleum and natural gas
		009	Iron ore
		010	Other metallic ore
		011	Non-metallic ore and quarrying
003	Basic industrial materials	012	Milled grain and flour
		018	Spinning
		019	Weaving and dyeing
		024	**Timber** (Moved from 004 to 003)
		026	**Other wooden products** (Moved from 004 to 003)
		027	Pulp and paper
		028	Printing and publishing
		029	Synthetic resins and fiber
		030	Basic industrial chemicals
		031	Chemical fertilizers and pesticides
		032	Drugs and medicine
		033	Other chemical products
		034	Refined petroleum and its products
		035	Plastic products
		036	Tires and tubes
		037	Other rubber products
		038	Cement and cement products
		039	Glass and glass products
		040	Other non-metallic mineral products
		041	Iron and steel
		042	Non-ferrous metal
		043	Metal products
00-a	**Wearing apparel and other made-up textile products**	020	**Knitting** (Separated from 004)
		021	**Wearing apparel** (Separated from 004)
		022	**Other made-up textile products** (Separated from 004)
00-b	**Other non-electrical consumption products for daily-use**	013	Fish products
		014	Slaughtering, meat products and dairy products
		015	Other food products
		016	Beverage
		017	Tobacco

TIIO2005 (Differences from TIIO2000 are shown in bold.)

15 Sector Classification		76 Sector Classification of the 2000 AIO Table	
Code	**Description**	**Code**	**Description**
005-e	**Other processed and assembled manufacturing products**	023	Leather and leather products
		025	Wooden furniture
005-a	**Computers and electronic equipment**	050	Electronic computing equipment
		051	Semiconductors and integrated circuits
		052	Other electronics and electronic products
005-b	**Automobiles**	055	Motor vehicles
		056	Motor cycles
005-c	**Industrial machinery**	044	**Boilers, engines and turbines** (Separated from 005)
		045	**General machinery** (Separated from 005)
		046	**Metal working machinery** (Separated from 005)
		047	**Specialized machinery** (Separated from 005)
005-d	**Household electrical appliance**	049	Television sets, radios, audios and communication equipment
		053	**Household electrical equipment** (Moved from 004)
		048	Heavy electrical equipment
		054	Lighting fixtures, batteries, wiring and others
		057	Shipbuilding
		058	Other transport equipment
		059	Precision machines
		060	**Other manufacturing products**
006	Electricity, gas and water supply	061	Electricity and gas
		062	Water supply
007	Construction	063	Building construction
		064	Other construction
008	Trade	065	Wholesale and retail trade
009	Transportation	066	Transportation
010	Other services	067	Telephone and telecommunication
		068	Finance and insurance
		069	Real estate
		070	Education and research
		071	Medical and health service
		072	Restaurant
		073	Hotel
		074	Other services
		075	Public administration
		076	Unclassified
Value Added		**Value Added and Final Demand Items**	
		001	Wages and salary
		002	Operating surplus
		003	Depreciation of fixed capital
		004	Indirect taxes less subsidies
Final Demand			
		001	Private consumption
		002	Government consumption
		003	Gross fixed capital formation
		004	Changes in stocks

Table 1: Industrial sector classification, TIIO2000 and TIIO2005

Valuation

The table is valued at producers' prices. This is because the benchmark tables of core/endogenous countries are also valued at producers' prices. The import matrix from the Rest of the World is valued at c.i.f. (cost-insurance-freight).

Compilation Procedure

We employ what we dubbed the 'split-in' method for compiling TIIO tables. That is, we start with the Asian International Input-Output Table as a basic framework (or control totals) for tabulation. Then, by using 'split ratios' derived from interregional input-output tables of core countries (China, Japan and Korea), the transaction matrices in the AIIO are subdivided by regions as specified above. This method is cost-effective because the AIIO is already balanced and internally-consistent and thus allows us to avoid the laborious adjustment procedure for harmonising data conflicts between different national input-output tables.

Concluding Remarks: Why do we Care about Sub-Regions?

The current MRIO framework regards a referred country as if it is a 'point' of transaction in the international production networks. A national economy, however, has a spatial dimension. It is rather unjustifiable to treat countries like China or Russia in the same manner as Singapore or Costa Rica in the input-output matrices. The argument is particularly relevant for the issue of regional development. For example, China is known to have been highly successful in building up strong economic linkages with neighbouring countries after the launch of the Reform and Open-door Policy in 1978. The benefit of economic globalisation was however not equally shared within the country; income disparity immediately enlarged, especially between coastal regions and inland regions, and it took some time for the positive impact from abroad to trickle down to inner China through domestic linkage effects. In this sense, the regional aspects are crucial in considering the process of economic development, at least for the spacious and less integrated economies like China and Russia.

The regional aspect also matters when we consider the impact of foreign direct investment. As a result of increasing relocation of production sites across borders, it is possible to envisage that one region in a country has stronger economic linkages with regions in foreign countries rather than with its own domestic neighbours. There is evidence to show that many manufacturing firms in Kyushu region have relocated their production sites to Korea's special economic zones after the recent devastating earthquake in Eastern Japan and the sharp appreciation of Japanese yen that followed. So, there is reason to assume that Kyushu region has stronger economic ties with Korea than the rest of Japan. Such a phenomenon can only be analysed with the help of detailed information of interregional input-output relations across countries.

Further reading

IDE-JETRO. (2007). 2000 Transnational Interregional Input-Output Table between China and Japan. IDE-JETRO.

Meng, B., & Qu, C. (2008). Application of the Input-Output Decomposition Technique to China's Regional Economies. Journal of Applied Regional Science , 13, 27-46.

Meng, B., Zhang, Y. X., Guo, J. M., & Fang, Y. (2012). China's Regional Economics and Value Chains: An Interregional Input-Output Analysis, IDE Discussion Paper No 359.

Miller, R., & Blair, P. (2009). Input-Output Analysis: Foundations and Extensions. 2nd edition. Cambridge, UK: Cambridge University Press.

WTO and IDE-JETRO. (2011). Trade Patterns and Global Value Chains in East Asia: From trade in goods to trade in tasks, WTO/IDE-JETRO.

Disclaimer

(1) The countries/regions referred to in this paper are presented in the alphabetical order of their regional codes, and it does not represent the author's judgment on the preference/ranking among them.

(2) The economic entities are frequently referred to as 'countries' in this paper, although some of them are not countries in the usual sense of the word but are officially 'customs territories'.

(3) Geographical and other groupings in this paper do not imply any expression of opinion by the author concerning the status of any country or territory, the delimitation of its frontiers or the rights or obligations of any sovereign states.

(4) The colours, boundaries, denominations, and classifications used in the maps which are featured in this paper do not imply any judgment of the legal or other status of any territory, nor any endorsement or acceptance of any boundary.

(5) Throughout this report, the Republic of Korea and the Separate Customs Territory of Taiwan, Penghu, Kinmen and Matsu are referred to as Korea and Taiwan, respectively.

Chapter 5: The World Input-Output Tables in the WIOD Database

Erik Dietzenbacher, Bart Los, and Marcel Timmer

Introduction: why WIOD?

If the amount of electronic products that German households import from China increases, what is the effect on the employment of low-skilled workers and on CO_2 emissions in Korea? This is a typical question reflecting that today's products and services are no longer produced within a single country. Whereas a product may list that it is 'made in China', its key components are often produced also in other parts of the world. In the last couple of decades, production processes have been sliced up more and more into ever smaller parts, each of which is carried out by a different producer. In addition, this fragmentation more and more crosses the borders of countries. The common viewpoint today is that products and services are made in global value chains. However, as pointed out by the European Commissioner for Trade, Karel De Gucht:[1] "though we are aware of the rising importance of global value chains, we have so far been unable to properly measure their size, nature and effect. This is because our current statistical apparatus does not capture the domestic activity contained in a traded good or service."

Economic and environmental policies are designed at a very detailed level of industries and products. At the same time, production processes are characterized by fragmentation leading to an interdependent structure, where everything depends on everything. The data that provide a description of such an interdependent production structure are given in supply and use tables (SUTs)

---

[1] Available at: http://trade.ec.europa.eu/doclib/docs/2012/april/tradoc_149337.pdf

and/or input-output tables (IOTs). Given the on-going trend in globalization, a database that is useful for policy-making and analysis should take *each* of the following three aspects into account. It must: (i) be *global*, (ii) cover *changes over time* in order to evaluate past developments, and (iii) include a variety of *socio-economic and environmental indicators*. Moreover, it is necessary to have all data in a coherent framework (e.g. using the same industry classification).

The WIOD project (World Input-Output Database: Construction and Applications) was set up to create such an all-encompassing database[2]. The project ran for three years (May 2009 – April 2012) and 11 international partners were involved[3]. The database provides the statistical apparatus that De Gucht was referring to. It allows for addressing issues related to fragmentation and socio-economic aspects (such as jobs or the creation of value added) as well as environmental aspects (such as energy use, various emissions to air, or the use of water). The database combines detailed information on national production activities and international trade data. For each country, tables are used that reflect how much of each of 59 products is produced and used by each of 35 industries. By linking these tables to trade data it is estimated, for example, how many dollars of Belgian fabricated metal products are used by the French transport equipment industry. This type of information is available in the WIOD database: for 40 countries (all 27 EU countries and 13 major other countries), plus estimates for the rest of the world; for the time period 1995-2007 (and estimates for 2008 and 2009); in current prices while "deflated" information is given in previous year's prices. It should be emphasized that all data in WIOD are obtained from official national statistics and are consistent with the National Accounts. The full database is publicly available and free of charge at: http://www.wiod.org/database/index.htm.

The rest of this chapter is structured as follows. Section 2 will be devoted to the principles underlying the database, i.e. the choices that have been made in its construction. Different (groups of) researchers make different choices, so that the world input-output tables (WIOTs) constructed by one (group of) researcher(s) differ from the WIOTs constructed by another[4]. A careful inspection of these principles (next to the coverage of the WIOTs) should help the user to make a choice between the available WIOTs. Section 3 will provide a very brief

---

[2] More information about the WIOD project can be found at: http://www.wiod.org/index.htm

[3] The participants are: University of Groningen, The Netherlands; Institute for Prospective Technological Studies, Seville, Spain; Wiener Institut für Internationale Wirtschafsvergleiche, Vienna, Austria; Zentrum für Europäische Wirtschaftsforschung, Mannheim, Germany; Österreichisches Institut für Wirtschaftsforschung, Vienna, Austria; Hochschule Konstanz, Germany; The Conference Board Europe, Brussels, Belgium; CPB Netherlands Bureau for Economic Policy Analysis, The Hague, The Netherlands; Institute of Communication and Computer Systems, Athens, Greece; Central Recherche SA, Paris, France; OECD, Paris, France.

[4] In the environmental and ecological literature, WIOTs are often termed multi-region input-output tables (or MRIOs).

description of the construction of the WIOTs in the WIOD database[5]. The last section will focus on a brief description of the full WIOD database.

The Underlying Principles and Choices

Constructing a large database like in the WIOD project implies that choices need to be made. Thinking about the options comes close to thinking about the fundamental principles underlying the construction of one's own database and, therefore, also the differences with other databases. Consequently, it is *not* the case that one database is better than another database. It may be better (or more appropriate) for answering some questions but not for other questions. For example, it is a choice to include as many countries as possible or to aim at confidence in the sources by limiting the number of countries (knowing that for certain countries the quality of the data is poor or that data are not official). That is, a choice between quantity and quality should be made. Another example is the choice how to deal with discrepancies that exist between the export and import values recorded in the national accounts statistics (NAS) and in the international trade statistics (ITS)[6]. One option is to take the absolute values from NAS and to assign these to countries-of-origin and countries-of-destination using shares obtained from ITS. Another option is to take the values from ITS and adapt the product-level exports and imports from NAS. The rest of this section will point out the choices that have been made in such instances, in constructing the WIOD database.

Using National Supply and Use Tables as the Starting Point for the Construction.

As building blocks, we have used national supply and use tables (SUTs). These are the core statistical sources from which National Statistical Institutes derive national input-output tables (IOTs). In IOTs it is assumed that each industry produces exactly one product. Consequently, the distinction between industry and product vanishes and the tables become square (or, in statistical parlance, symmetric). SUTs on the other hand are usually non-square and allow for secondary production, which makes them better reflect 'reality'. The supply table provides information on how much of each product is produced by each domestic industry. The use table indicates the use of each product (combining domestically produced and imported products) by each of the industries and final use categories (e.g. consumption by households and government, investments, and gross exports)[7]. Both supply tables and use tables are thus of the product-by-industry dimension. Therefore, linking SUTs with international trade data (which are product based) and with socio-economic and environmental data (which are

---

[5] A detailed description of the contents, sources and methods for the WIOD database can be found in Timmer (ed., 2012).
[6] See the discussion in Guo *et al.* (2009).
[7] See Miller and Blair (2009) for an elaborate introduction to IOTs and SUTs, and on the derivation of IOTs from SUTs.

mainly industry-based) becomes more natural (i.e. does not require a transformation of the source data).

Using National Accounts as a Benchmark

Typically, SUTs are only available for a limited set of years (e.g. every 5 years) and once released by the national statistical institute revisions are rare. National Accounts on the other hand are usually revised. This is because statistical systems develop, new methodologies and accounting rules are used, classification schemes change and new data become available. Occasionally, revisions are also carried through to ensure consistency and comparability over time. These revisions can be substantial, especially at a detailed industry level, implying discrepancies between information from the latest version of the National Accounts for a certain year and the published SUT for that year. The SUTs in WIOD are therefore benchmarked on the National Accounts. So, any revision of the National Accounts leads to an adaptation of the national SUTs to make them match.

Constructing Time Series

One of our aims was to arrive at a time series of WIOTs. Because national SUTs are only infrequently available and are often not harmonized over time, they have been benchmarked on consistent time series from the national accounts statistics (NAS). Time series for (gross) output and value added by industry, total imports and total exports and final use by use category were taken from the NAS. These data were used as constraints when generating time series of SUTs with the so-called SUT-RAS method. This method is akin to the well-known RAS-technique (a bi-proportional updating method for IOTs). This technique has been adapted for updating SUTs and has been shown to outperform other methods for generating time series of SUTs[8].

Time series of SUTs have been derived for two price concepts: basic prices and purchasers' prices. Basic prices reflect all costs borne by the producer, whereas purchasers' prices reflect the amounts paid by the purchaser. Supply tables are always in basic prices and often have additional information on margins and net taxes by product. Use tables as available from public data sources are typically in purchasers' prices and have to be transformed to basic prices within the construction procedures. The difference between the two types of use tables is given in the so-called valuation matrices with the trade and transportation margins and the net taxes, which had to be estimated.

Using Only Publicly Available Data

Within WIOD, the choice was made that the data used in the project should be publicly available. This ensures that users of the WIOTs are able to trace the construction process and are able to derive alternative tables by making a

---

[8] See Temurshoev and Timmer (2011).

different set of assumptions. Moreover, officially published data are more reliable because thorough checking and validation procedures have been adopted by National Statistical Institutes (when compared to data generated on an ad-hoc basis for specific research purposes).

Improved Allocation of Imports of Goods

In the process of construction, the national SUTs have been combined with information from international trade statistics to construct what we call international SUTs. Recall that use tables include both domestically produced and imported products. They have been split into use of domestic products and use of foreign products first, and in a second stage use of foreign products is split according to country of origin. The standard assumption in most databases is to apply import proportionality (where the same, fixed percentage of total use of a product is assumed to be imported, irrespective of its purchaser). For example, if 40% of the total purchases (by industries and final users) of rubber and plastic products are imported, the same 40% applies to each and every industry-of-use and final demand category. That is, 40% of rubber and plastics used by the transport equipment manufacturing industry (or any other industry) are imported, 40% of the household consumption of rubber and plastics are imported, and the same applies to investment and government consumption.

For the imports of goods we have developed an estimation method that does not rely on this standard import proportionality assumption. The UN COMTRADE database provides information on bilateral flows of goods (at the HS6-digit level) for about 5000 products. For each of these products the shares of its imports that went to 'intermediate consumption, to 'final consumption', and to 'capital goods' were determined (i.e. modifying the end-use categories in the Broad Economic Categories classification as provided by the UN). Imports have been allocated across end-use categories in the following way. The share of any end-use category (intermediates, final consumption, or investment) was used to split up total imports for each of the 59 products in the WIOD classification *across* the three end-use categories. W*ithin* each end-use category, the allocation was based on the proportionality assumption. A similar procedure was used to split the imports table according to country of origin. Unlike the case for the standard proportionality assumption, country import shares differ across end-use categories (but not within these categories).

Additional Choices

Given the types of application that we had in mind for the WIOD database, it is important also to have detailed information on the trade in services. For services trade, however, no standardized database on bilateral flows exists. The data have been collected from various sources (including OECD, Eurostat, IMF and WTO), checked for consistency and integrated into a bilateral services trade database.

For some applications it is important to have data in constant prices. Therefore WIOTs in previous years' prices have been constructed based on exporters' gross output deflators.

One of our aims was to link the WIOTs to satellite accounts that provide data at the same industry level. The socio-economic accounts focus on the inputs of production factors and the environmental accounts list requirements for and effects of production (e.g. energy use and emissions, respectively)[9].

Construction of the WIOTs in a Nutshell

The construction of the WIOTs comprised four steps. First, data were collected and harmonized, after which time series of national SUTs were constructed in the second step. Third, these national SUTs were linked across countries through detailed bilateral international trade statistics. This yielded so-called international SUTs. Fourth, these international SUTs were then used to construct the symmetric WIOTs that are of the industry-by-industry type.

In the first step, three types of publicly available data have been used. These are national accounts statistics, SUTs, and international trade statistics. The data have been harmonized in terms of industry- and product-classifications, both across time and across countries. The WIOD classification lists 59 products and 35 industries based on the CPA and NACE rev 1 (ISIC rev 2) classifications. To arrive at a common classification, correspondence tables have been made for each national SUT, bridging the level of detail in the country to the WIOD classification. This involved aggregation and sometimes disaggregation based on additional detailed data. While for most European countries this was relatively straightforward, tables for non-EU countries proved more difficult. National SUTs have also been checked for consistencies and adjusted to common concepts (e.g. regarding the treatment of FISIM, financial intermediation services indirectly measured, and purchases abroad)[10]. Finally, because national SUTs are in national currencies, official exchange rates from the IMF were used to have all data listed in dollars.

The second step led to a time series of SUTs, as described in the previous section. In the third step, the national SUTs were combined with information from international trade statistics to construct what we have called international SUTs. Recall that use tables include both domestically produced and imported products. They have been split according to domestic or foreign origin first and then according to country of origin based on import shares. As described in the previous section, for goods we have adopted an improved estimation method and for services a bilateral service trade database was constructed.

In the final step, the international SUTs for all countries were combined into a world SUT first. That is, they were stacked and reordered to resemble a standard supply and use table. Subsequently, this world SUT was transformed into the WIOT. Recall that IOTs are symmetric and can be of the product-by-product type (describing the amount of products needed to produce a particular good or service) or of the industry-by-industry type (describing the flow of goods and services from one industry to another). The choice between the two types

---

[9] See Erumban *et al*. (2012a) and Genty *et al*. (2012) for a detailed description.
[10] The adjustments and harmonization are described in more detail on a country-by-country basis in Erumban *et al*. (2012b).

depends on the research questions. Our choice was to derive industry-by-industry IOTs because many applications were foreseen that would use industry level data (e.g. value added, employment, energy use, emissions) and because industry-by-industry tables link better with national account statistics. An IOT is a construct on the basis of a SUT at basic prices using additional assumptions concerning technology. We have used the so-called 'fixed product sales structure' assumption, stating that each product has its own specific sales structure (reflecting the proportions of the output of the product in which it is sold to the respective intermediate and final users) irrespective of the industry where it is produced. This assumption is most widely used, because it is more realistic, because it requires a relatively simple mechanical procedure, and because it does not generate any negatives in the IOT that would require manual rebalancing. The IOT that has been constructed in this way is an inter-country IOT for forty countries with matrices for imports from RoW and a vector of gross exports to RoW. Various modeling purposes and analyses, however, require a full WIOT. Therefore the Rest of the World has been added as a single region, as a proxy for all other countries in the world. The estimation was based on totals for industry output and final use categories from the UN National Accounts, assuming an input structure equal to that of an average emerging country (for which the structures of Brazil, Russia, India, China, Indonesia, and Mexico have been taken into account).

The Contents of the Full World Input-Output Database

The World Input-Output Database covers 27 EU countries and 13 other major countries in the world and provides annual data for the period from 1995 to 2009. The countries are listed in Table 1 and were selected on the basis of the quality and the public availability of their data in combination with their economic importance (together they account for approximately 85% of world GDP). All SUTs are of the format 35 industries by 59 products and all IOTs are 35 industries by 35 industries. All data are downloadable at http://www.wiod.org/database/index.htm. The database includes the following information.

World Tables
- International SUTs in current and previous year's prices, with use split into domestically produced products and imported products by country of origin
- WIOTs in current prices and previous year's prices
- Interregional IOTs for 6 regions (see Table 1)

National Tables
- National SUTs in current and previous year's prices
- National IOTs in current prices

Socio-Economic Accounts
- Industry output, value added, in current and constant prices (1995 = 100)
- Capital stock, investments

- Wages and employment (in hours) by skill type (low-, medium- and high-skilled)

Environmental Accounts
- Gross energy use by sector (i.e. industries and households) and energy commodity
- Emissions relevant energy use by sector and energy commodity
- CO_2 emissions modeled by sector and energy commodity
- Emissions to air by sector and pollutant
- Land use, materials use and water use by type and sector

Table 1: WIOD countries and their regional aggregation

Euro-zone	Non-Euro EU	NAFTA	China	East Asia	BRIIAT
Austria Belgium Cyprus Estonia Finland France Germany Greece Ireland Italy Luxembourg Malta Netherlands Portugal Slovakia Slovenia Spain	Bulgaria Czech Rep. Denmark Hungary Latvia Lithuania Poland Romania Sweden UK	Canada Mexico USA	China	Japan Korea Taiwan	Brazil Russia India Indonesia Australia Turkey

Further Reading

Erumban, A., Gouma, R., de Vries, G., de Vries, K., & Timmer, M. P. (2012). *WIOD Socio-Economic Accounts (SEA): Sources and Methods.* Retrieved 2012 from WIOD:
http://www.wiod.org/publications/source_docs/SEA_Sources.pdf

Erumban, A., Gouma, R., de Vries, G., K., d. V., & Timmer, M. P. (2012). *Sources for National Supply and Use Table Input Files.* Retrieved 2012 from WIOD:
http://www.wiod.org/publications/source_docs/SUT_Input_Sources.pdf

Genty, A. I., Neuwahl, A., & Neuwahl, F. (2012). *Final Database of Environmental Satellite Accounts: Technical Report on their Compilation.* Retrieved 2012 from WIOD:

http://www.wiod.org/publications/source_docs/Environmental_Sources.pdf

Guo, D., Webb, C., & Yamano, N. (2009). *Towards harmonised bilateral trade data for inter-country input-output analyses: statistical issues, OECD STI Working Papers 2009/4.* OECD. OECD, Directorate for Science, Technology and Industry.

Miller, R. E., & Blair, P. D. (2009). *Input-Output Analysis: Foundations and Extensions, second edition* . Cambridge, UK: Cambridge University Press.

Temurshoev, U., & Timmer, M. P. (2011). Joint estimation of supply and use tables. *Papers in Regional Science* (90), 863-882.

Timmer, M. P. (Ed.). (2012). *The World Input-Output Database (WIOD): Contents, Sources and Methods, WIOD Working paper no 10.* WIOD.

Acknowledgements

The WIOD (World Input-Output Database: Construction and Applications) project was funded by the European Commission, Research Directorate General as part of the 7th Framework Programme, Theme 8: Socio-Economic Sciences and Humanities. Grant Agreement no: 225 281.

Chapter 6: EXIOBASE – A Detailed Multi-Regional Supply and Use Table with Environmental Extensions

Arnold Tukker

Introduction

The EXIOPOL (A New Environmental Accounting Framework Using Externality Data and Input-Output Tools for Policy Analysis) was an EU funded project executed between 2007 and 2011 that had two main goals. One part of the project aimed at improving insights into external costs of environmental pressures. The other part, central in this paper, tried to overcome significant limitations in existing data sources in the field of Multi-regional Environmentally Extended Supply and Use Tables (MR EE SUT) and Input Output Tables (IOT). National Statistical Institutes (NSIs) provide SUT and IOT for single countries, without trade links. Sector and product detail is not as good as it ought to be. Environmental extensions are often lacking or include only a few types of emissions and primary resource uses. There is little or no harmonization of sector and product classifications across different countries. It is therefore difficult to assess the extent to which a country induces environmental impacts abroad via trade, let alone trends therein. Trade-linked tables are also essential for analyzing the effects of sustainability measures taken in Europe on Europe's economic competitiveness. From a theoretical viewpoint, the MR EE IO approach is the best way of taking trade into account, but existing studies tend to be aggregated at sector and regional level and to focus on a fairly small number of environmental extensions. Figure 1 shows a stylized example of a 4 country MR EE SUT.

The EE IO work in EXIOPOL sought to make a crucial advance in finding a solution to this problem. The project's aim was really to leapfrog: it would give the EU a fully-fledged, detailed, transparent, public, global MR EE IO database with externalities, allowing for numerous types of analyses for policy support purposes[1].

This paper discusses the following topics related to the EXIOPOL project. First, we present the construction of the database, discussing the main data sources and estimation/construction methods (section 2). We then give some illustrative results (section 3) and end with conclusions (section 4).

Figure 1: Stylized MR EE IOT. Red: individual country IOT (with Z being intermediate demand, Y final demand, W, g, C, L the value added block (profits, capital expenditure, labor). Green: bilateral trade matrices (with Z for intermediate use, and Y for final use). Grey: extensions (NAMEA = emissions, other blocks: various forms of resource extraction)

	Industries				$Y_{*,A}$	$Y_{*,B}$	$Y_{*,C}$	$Y_{*,D}$	q
Products	$Z_{A,A}$	$Z_{A,B}$	$Z_{A,C}$	$Z_{A,D}$	$Y_{A,A}$	$Y_{A,B}$	$Y_{A,C}$	$Y_{A,D}$	q_A
	$Z_{B,A}$	$Z_{B,B}$	$Z_{B,C}$	$Z_{B,D}$	$Y_{B,A}$	$Y_{B,B}$	$Y_{B,C}$	$Y_{B,D}$	q_D
	$Z_{C,A}$	$Z_{C,B}$	$Z_{C,C}$	$Z_{C,D}$	$Y_{C,A}$	$Y_{C,B}$	$Y_{C,C}$	$Y_{C,D}$	q_C
	$Z_{D,A}$	$Z_{D,B}$	$Z_{D,C}$	$Z_{D,D}$	$Y_{D,A}$	$Y_{D,B}$	$Y_{D,C}$	$Y_{D,D}$	q_D
W	W_A	W_B	W_C	W_D					
g	g_A	g_B	g_C	g_D					
C & L	$Capital_A$	C_B	C_C	C_D					
	$Labor_A$	L_B	L_C	L_D					
Environ Ext	$NAMEA_A$	$NAMEA_B$	$NAMEA_C$	$NAMEA_D$					
	$Agric_A$	$Agric_B$	$Agric_C$	$Agric_D$					
	$Energy_A$	$Energy_B$	$Energy_C$	$Energy_D$					
	$Metal_A$	$Metal_B$	$Metal_C$	$Metal_D$					
	$Mineral_A$	$Mineral_B$	$Mineral_C$	$Mineral_D$					
	$Land_A$	$Land_B$	$Land_C$	$Land_D$					

---

[1] Tukker, A., Poliakov, E., Heijungs, R., Hawkins, T., Neuwahl, F., Rueda-Cantuche, J., Giljum, S., Moll, S., Oosterhaven, J. & Bouwmeester, M. (2009). Towards a global multi-regional environmentally extended input-output database. Ecological Economics 68 (7), 1929-1937.

Data Case Construction: Main Data Sources and Estimation Methodology

Introduction

The following main steps were taken in constructing the database:

- Harmonizing and detailing SUT
- Harmonizing and estimating environmental extensions
- Linking the country SUT via trade to an MR EE SUT
- Transforming the MR EE SUT into various types of MR EE IOT.

We will discuss our approach in brief in the next sections, ending with a description of the full database. At the start of the project the most recent data available was for 2000, therefore this was chosen as the base year.

Harmonizing and Detailing SUT

As indicated, many countries publish SUT and/or IOT. They are usually based on a common UN classification for sectors and products (the so-called ISIC and CPA classifications). Yet, it appears that in most cases countries choose various levels of product and sector detail, and tend to cluster sectors and products differently. The European Statistical office (EUROSTAT) has harmonized SUT and IOT for most European countries at a level of 60 products and sectors, but for other countries detail can range from a few dozen sectors (as for example is the case for Russia) to around 500 sectors (as is the case for the US and Japan). From an environmental perspective, it was very unfortunate that most countries distinguish just one agricultural sector, one power generation sector, and one transport sector – where the impacts of meat production for example are much higher than those of cereal production, and in the case of power production the impacts of coal are much higher than those of gas.

   A major challenge in EXIOPOL was hence harmonizing sector and product classifications across countries, and creating more detail in sectors and products that have a high environmental relevance. Various technical issues in the original tables had to be solved too. EXIOPOL used a two-step procedure for this.

1. First, all SUT and IOT as sourced from statistical offices were transformed to a standard set of SUT in original classification and currency. This process resulted in the following tables for each country.

    a. Supply and Use Tables in basic prices; including valuation layers for the Use table. If these valuation layers are added to the Use table in basic prices this results in the Use table in purchaser prices (in which statistical offices usually report the Use table).

    b. A split of the Use table into a domestic Use part (i.e. use of products *made within* the country) and an import Use part (i.e.

use of products *made in other* countries, which hence were imported).

2. After this, the SUT were harmonized and mapped on the EXIOBASE classification, which for most countries implied a detailing procedure. Multiplied by Market exchange rates, the SUT then could be expressed in Euro, the currency used for EXIOBASE. This roughly worked as follows.

    a. A correspondence table between the original classification and the EXIOBASE classification was made; usually 1 original product or sector corresponded with 1 or more EXIOBASE products or sectors. For countries with detailed tables, we often could aggregate products and sectors to the EXIOBASE classification.
    b. From other sources, usually information was available on the total product output and the total industry turnover in a much higher detail than in the SUT[2]. This allowed estimating the split of row- and column totals in the SUT, and hence gave insight into the total industry turnover in EXIOBASE classification, as well as the total product output in EXIOBASE classification. However, this obviously did not yet help to analyse which (more detailed) industry would use the (more detailed) products – the inner part of the SUT matrix still was not fully filled.
    c. Again from other sources (e.g. countries with detailed tables), estimates could be obtained about the detailed use and supply of products per industry. These co-efficients then were used to estimate the product input and output at the more detailed level. Since this led to a not yet fully consistent result, a balancing program was used to iron out differences.

Harmonizing and Estimating Extensions

The environmental extensions include the use of primary resources (materials, water, energy and land) and emissions. It concerns in essence an emission and primary resource use database by country, but with the special feature that it provides data specific for each economic sector present in the SUT. Many countries have emission inventories, but often they are not linked to the industry classifications used by statistical offices. We harmonized and estimated extensions as follows.

1. Resource extraction data. One of the project partners, SERI, had already compiled a major database with data on extraction of resources such as

---

[2] For instance, EUROSTAT has a database called 'PRODCOM' that registers total production of products in the EU at a high level of detail.

oil, gas, minerals, agricultural products etc. per country[3]. Allocating such data to the right sectors was quite straightforward: coal obviously would correspond to the sector 'coal mining', wheat would correspond to the 'wheat production' sector, etc.

2. Emissions. Energy-related emissions are by far the most important. For each country, the International Energy Agency has a database indicating which energy carrier (e.g. gas, coal, oil) is used in which quantities by which economic sectors. We had to reclassify this IEA database to the EXIOBASE sectors, but after that the product of energy flows and country- and sector specific emission factors (sourced from authoritative sources such as the IPCC) could give robust estimates of emissions, calculated in a consistent manner across countries. For non-energy emissions, we also gathered physical data on activities and multiplied them with emission factors. For instance, N_2O emissions depend on nitrogen fertilizer use in agriculture. We hence gathered data on fertilizer use by sector, and multiplied this with the appropriate emission factors[4].

3. Land use data were obtained from the UN's Food and Agriculture Organisation (FAO), and could be allocated directly to the agricultural and forestry activities with the information present in that database. EXIOPOL did neglect direct land use by industrial sites, human settlements and other infrastructure, but this is just a small part of land use.

4. Water use mainly takes place in agriculture. Using various databases in combination with a global model, the water use per agricultural crop per country could be estimated. Other water extractions, particularly by industry and for cooling water, are more difficult to estimate – various data sources and assumptions were used here.

In this way, a list of hundreds of extensions by country and sector was obtained. This is far too detailed for practical use, and the project hence made use of various well-known indicator systems to present data in a more aggregated way. It concerns for instance:

- various environmental themes from Life Cycle Impact Assessment, most notably Global Warming Potential (GWP), Ozone Depletion Potential (ODP), Photochemical Oxidant Creation Potential (POCP), acidification and eutrophication;

---

[3] Such data in turn were based on major databases such as FAOSTAT for agricultural products, USGS for minerals and metals, and the IEA database for oil, gas and other fossil fuels.

[4] Pulles, T., M van het Bolscher, R. Brand and Antoon Visschedijk (2007). Assessment of Global Emissions from Fuel Combustion in the Final Decades ofthe 20th Century. Application fo the Emission Inventory Model TEAM. TNO Report 2007-A-R0132/B, TNO Built Environment and Geosciences, Apeldoorn, Netherlands

- various material flow-based indicators, which usually add up all resource extractions to a 'Total Material Requirement'.

We calculated a proxy for the 'Ecological Footprint' by calculating the actual land use related to making products, plus the 'virtual land use' that in theory is needed to compensate for energy use / greenhouse gas emissions.

Finally, for all emissions and specific to country and industry sector, rough estimates were made for the economic damage caused, most notably health effects (for instance, the value of life years lost).

Linking the country SUT via trade to an MR EE SUT

At this point we have for each country an EE SUT available in the same, harmonized classification and format. One of the EXIOPOL partners, the University of Groningen developed a method to link these tables via trade[5]. The main problem is that while SUT have imports and exports data in their tables, they do not give information about the countries of imports nor destination. Fortunately there are trade statistic databases, such as the UN COMTRADE database. This database is very detailed, but not in the classification used in SUT. Further, being databases made by different compilers, total imports and exports in COMTRADE usually are not entirely consistent with imports and exports in the SUT[6]. Hence, first a correspondence matrix was made aggregating UN COMTRADE to the EIOXBASE classification. We then calculated by product for all imports for each country the shares of origin of the imports. These shares were then used to make a first estimate of country of origin of the imports in each SUT. We did so for all 43 countries. This resulted for all countries into an economic value of say, wheat, imported from Canada. The sum of these imports in theory should be equal to the exports of wheat reported in the SUT for Canada[7]. Unfortunately, SUT being independently compiled by each country, there is usually a mismatch. Indeed, the total imports of wheat as reported by all countries in their SUT is unlikely to match exactly all the exports of wheat in their SUT![8].

---

[5] Bouwmeester, M.C. (2011). Algorithm applied on the full EXIOPOL SUT Data Set and documentation provided. DIII.4.a-4, RU Groningen. Groningen, Netherlands
[6] Another problem in UN COMTRADE is the 'mirror statistics puzzle'. UN COMTRADE simply stores what countries report. It can hence be that country A reports other imports from country B, as that country B reports to export to country A. For this reason, we used a version of UN COMTRADE harmonized already by other researchers. For more detail, see former note.
[7] Of course there are more than 43 countries in the world. The remaining countries in the world were grouped together in a 'Rest of world (RoW)' responsible for just over 5% of the global GDP. Also the imports and exports of the RoW were estimated so that also imports from other than the 43 countries in our database were included.
[8] This mismatch at global scale has the following reason. Where SUT by country must be by definition consistent, adhering to identities that supply=use, this does not work for all SUT combined since it is not possible for individual statistical offices to analyse consistency at global scale. Since it is unlikely that the 'Earth trades with aliens', the only solution is scaling and adapting imports and exports. It has to be noted that imports are

We hence applied a method that scaled for each product the total imports reported in country SUT so that they matched total exports reported in country SUT. We further allowed that initial estimates of country of origin of imports could be adjusted. This then led to consistent and balanced trade matrices and SUT, as close as possible to the original SUT and import and export values by product.

Transforming the MR EE SUT into various types of MR EE IOT

Most analytical applications and models used (e.g. Computable General Equilibrium model) are based on IOTs rather than SUT. Transforming SUT to IOT is a standard procedure described in various textbooks[9]. Figure 2 gives the procedure in a flow chart. EXIOPOL produced from its MR EE SUT an industry-by-industry MR EE IOT using the so-called Industry-technology assumption (ITA). The ITA assumes that inputs are required in proportion to output and the proportions are the same for an industry's primary and secondary products. An alternative approach, the Commodity Technology Assumption (CTA) assumes that the same products need the same product inputs in proportion to product outputs regardless which industry produces the product. The CTA is preferable theoretically, but it has the disadvantage of producing negative numbers. Due to this problem, many users apply the ITA despite the various theoretical issues.

Figure 2: Simplified input-output framework[10]

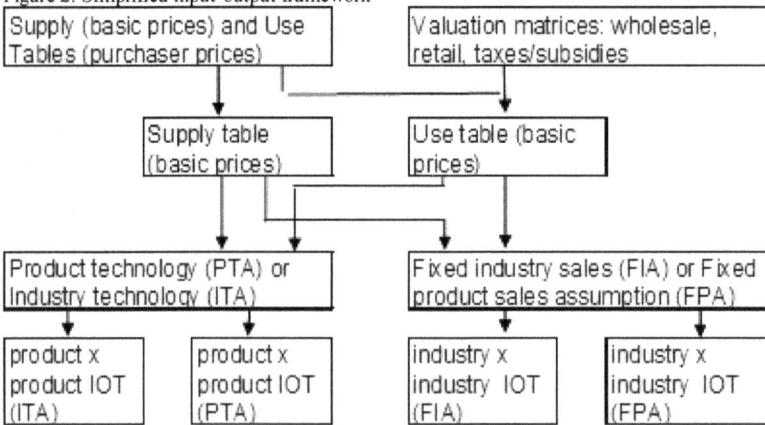

| Supply (basic prices) and Use Tables (purchaser prices) | Valuation matrices: wholesale, retail, taxes/subsidies |

| Supply table (basic prices) | Use table (basic prices) |

| Product technology (PTA) or Industry technology (ITA) | Fixed industry sales (FIA) or Fixed product sales assumption (FPA) |

| product x product IOT (ITA) | product x product IOT (PTA) | industry x industry IOT (FIA) | industry x industry IOT (FPA) |

---

usually including insurance and freight costs and exports without, our trade linking routine took these differences into account.

[9] Miller, R. and P.r Blair (2009). Input Output Analysis: Foundations and Extensions. 2nd Edition. Cambridge University Press, Cambridge, UK; Ten Raa, T (2005). The Economics of Input-Output Analysis, Cambridge University Press

[10] Rueda Cantuche, J., J. Beutel, F. Neuwahl, A. Löschl, and I. Mongelli (2007). A Symmetric Input Output Table for EU27: Latest Progress. Paper for the 16th IIOA Conference, Istanbul, July 2007, available from IPTS, Seville, Spain

The Result: EXIOBASE

Overview of the database

The result of the former steps is visualized in Figure 3: the EXIOBASE database. In essence it consists of three blocks.

Block 1: a database for the single country environmentally extended supply-use data. Into this block all data from the harmonization steps are imported and EE SUT for countries are created.

Block 2: the storage of the international supply-use table (interlinked supply-use table or MR EE SUT).

Block 3: the storage of the international input-output table (interlinked country input-output table or MR EE IOT).

Between block 1 and 2 the EE SUT are linked to a global MR EE SUT via an automated routine, using the method described in the section above. Since most analytical applications use IOT rather than SUT, another script creates in Block 3 the MR EE IOT (both of industry by industry and product by product MR EE IOTs). The MR EE SUT and MR EE IOTs that are available have the following characteristics.

- Covering 43 countries (95% of the global economy) and a Rest of World (the other 150+ countries in the world combined)
- Distinguishing 129 industry sectors and products
- Covering 30 emitted substances and 80 resources as extensions by industry.
- Full trade matrices: insight is given into which product from which country is exported to which industry sector in another country.

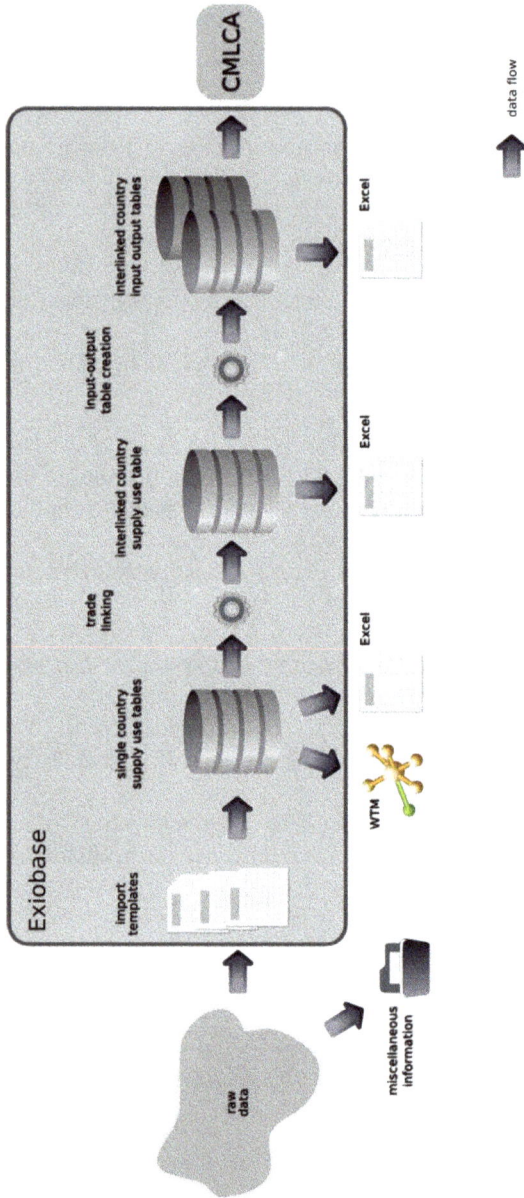

Figure 3: The EXIOBASE system. Raw data are transformed and harmonized into single country EE SUT. Single country EE SUT are transformed into a global MR EE SUT, and after that, to MR EE IOT. The MR EE IOT can be analysed with other software, such as CMLCA (Life cycle assessment software developed by EXIOPOL partner CML, Leiden University). EE SUT, the MR EE SUT and MR EE IOT are externally available in various formats, including Excel

Some illustrative results

As explained elsewhere in this book, an EE IOT can be used easily for calculation of impacts of final consumption. An MR EE IOT is in fact nothing more nor less than an EE IOT of the world. Hence, if one applies the standard calculation method for assessing impacts of final demand used for country EE IOT on a global, MR EE IOT, one now can assess impacts of final demand as well, but now attributed to sector and country of origin. In essence the MR EE IOT allows the calculation of which sectors in which country contribute which part of added value making up the full value of the finally sold product. With the emissions and resource extraction per unit of added value per sector and country also given in the MR EE IOT, the total impacts along the global value chain can be estimated too[11].

Using this calculation approach, the impacts related to final European demands were calculated with EXIOBASE and split between impacts in Europe and impacts abroad. Table 1 shows that particularly with regard to land use, water extraction and material use Europe depends for a large part on other countries.

Table 1: Impacts per capita related to EU27 final demand, 2000, as well as impacts related to EU27 imports and exports per capita. For instance, the final consumption of each European citizen needed 17 tons of primary material extraction (material extraction used). Of this, 6,5 tons were provided by non EU countries. This is much higher than the primary material extraction embedded in Europe's exports, which are 2.6 tons per capita.

Impact type	Unit	Final demand	Import	Export
External costs	Euro	1173	90	112
Land footprint	ha	1.7	1.0	0.1
Net energy use	GJ	116	29	21
Water consumption blue	m3	767	335	75
Water consumption green	m3	4446	2301	367
Domestic extraction used	ton	17.0	6.5	2.6
Unused domestic extraction	ton	13.8	4.5	1.8
Acidification	kg SO2-eq.	68.0	14.8	7.3
Eutrophication	kg PO4–eq.	9.1	2.0	0.9
Global warming potential	ton CO2-eq.	12.4	2.3	1.6

* Assuming an EU27 population of 483 Million in 2000 (figure taken from the Eurostat website, http://epp.eurostat.ec.europa.eu/cache/ITY_OFFPUB/KS-QA-09-031/EN/KS-QA-09-031-EN.PDF accessed 2 September 2011).

Conclusions

The EXIOPOL project created a detailed, global, multiregional environmentally extended supply and use table called EXIOBASE. The MR EE SUT and MR EE IOTs that are available have the following characteristics:

---

[11] In mathematical terms, the following formula applies: $x^E = Dx = D(I-A)^{-1}f$, where f = final demand, A is the matrix of direct input co-efficients, D is the matrix of direct impact coefficients, and x^E is the total requirement of environmental factors.

- Covering 43 countries (95 percent of the global economy) and a Rest of World (the other 150+ countries in the world combined);
- Distinguishing 129 industry sectors and products;
- Covering 30 emitted substances and 80 resources as extensions by industry; and
- Full trade matrices: insight is given into which product from which country is exported to which industry sector in another country.

Compared to other global MR SUT or IOT, the database is unique because of its global coverage, its detail in sectors, and broad coverage of emissions and other impacts. It is hence very well suited for environmental analysis. Compared to some other projects in the field of global MR SUT/IOT, EXIOBASE has the drawback that it does not yet contain time series – something envisaged to be developed in follow-up projects. The database is available via www.exiobase.eu in two forms: a free, aggregated version and a not for profit licensed detailed version. Static analyses are possible with other software, such as CML's CMLCA program, or Groningen University's IRIOS program. The fees are to allow for making updates of the database without further EU support.

Further reading

Eurostat. (2001). *Economy-wide Material Flow Accounts and Derived Indicators. A methodological guide.* Luxembourg: Eurostat, European Commission, Office for Official Publications of the European Communities.

Eurostat. (2008). *Eurostat Manual of Supply, Use and Input-Output Tables.* Luxembourg: Eurostat.

Lutter, S., Acosta, J., Kuenen, J., Giljum, S., Wittmer, D., & Pulles, T. (2011). *Documentation (Technical Report) of data sets compilation for environmental extensions. EXIOPOL Deliverables DIII.2b-2 & DIII.3.b-2.* Milan: FEEM, Italy.

Wackernagel, M., Monfreda, C., Moran, D., Wermer, P., Goldfinger, S., Deumling, D., et al. (2005). *National Footprint and Biocapacity Accounts 2005: The underlying calculation method.* Oakland: Global Footprint Network.

Acknowledgements

Thanks go obviously to the whole EXIOPOL team for performing the project on which basis this paper was written. A special word of thanks is for Arjan de Koning who did all calculations in earlier reports leading to the results presented in section 3. EXIOPOL was funded by the 6th Framework Program of the European Commission (EC). This paper does not represent any official position or endorsement of the funding organization.

Chapter 7: The Eora MRIO

Dan Moran

Almost as soon as Leontief began IO analysis he had envisioned a global information system, a complete and detailed schematic of the world economy. That goal has remained stubbornly out of reach until only recent years. The Eora MRIO is one project to build such a table. We aimed to build a large, high-resolution global MRIO, and to do so using automation as much as possible.

Background

As international trade and multinational production become ever more important there is a need for economic analysis that doesn't hit walls at national borders. Multi-region IO tables such as the GTAP, EXIOBAS, GRAM, and others described in this book respond to that need.

In particular, IO analysis is currently in the midst of a renaissance in its use for environmental applications. Its ability to trace out the supply chains linking consumers to the impacts they ultimately drive makes it a powerful tool for guiding policymaking. So far most environmental policy is based on a principle of producer responsibility, the idea that the company directly causing an impact is responsible for it. But there is growing recognition that the consumers of those implicated products – who ultimately benefit from that environmental harm – should also play a role in reducing their footprint. This idea is called the consumer responsibility principle. China has suggested this principle be used when setting the next round of GHG reduction targets. Up to 40% of China's emissions are due to the production of export goods. Complementarily, the UK has recently been disappointed to find that reductions in its territorial GHG emissions have been more than offset by increases in its carbon footprint embodied in imports.

Successful sustainability policy will consider both producer and consumer responsibilities, and will need to address different challenges at the production, trade, and consumption points along the supply chain. MRIO schematics of the world economy allow us to trace supply chains and differentiate policy at different stages of the supply chain to achieve a coordinated effect. Parts IV and V of this book describe a number of case studies and applications where global MRIO databases can provide valuable new insights based on their detailed accounts of international flows.

The Eora MRIO project has several goals:

- Preserve all original data from national statistical agencies in their original high-resolution classification (integrity of original data)
- Provide full country coverage (180+ countries)
- Provide a continuous timeseries from at least 1990
- Provide uncertainty information for all elements
-

In developing Eora we hoped to significantly advance the state of the art in each of these aspects, and furthermore we hoped to accomplish this at a low cost (<$1 million) through the extensive use of software (Table 1). The table covers 187 countries with 25-400 sectors per country (for a total of 15,000 sectors) and five valuation sheets, over the period 1990-2011. Eora captures >99.99% of global trade. Every data point is accompanied by an estimate of its standard deviation.

Table 1 – The Eora MRIO achieves a high level of coverage and detail

Goal	Realization
Country coverage	187 individual countries
Sector coverage	15,909 sectors[1]
International trade	By destination country and sector
Environmental indicator coverage	35 indicators[2] with >1700 single indicators
Cost	US$ 500,000 development, US$ 250,000 p.a., ongoing
Continuity	Annual tables 1990-2011
Timeliness	1-2 years prior to current year
Reliability and uncertainty information	Standard deviation estimate for every value

---

[1] 187 single countries at 25-500 sectors totalling 14787 sectors, 5 valuation sheets, 20 years, makes in total more than 20 billion transactions.

[2] Energy, CO_2, CH_4, N_2O, HFC-125, HFC-134a, HFC-143a, HFC-152a, HFC-227ea, HFC-23, HFC-236fa, HFC-245fa, HFC-32, HFC-365mfc, HFC-43-10-mee, C_2F_6, C_3F_8, C_4F_{10}, C_5F_{12}, C_6F_{14}, C_7F_{16}, CF_4, c-C_4F_8, SF_6, HANPP, CO, NO_x, NMVOC, NH_3, SO_2, HC, HCFC-141b HCFC-142b, Ecological Footprint, Water Footprint, and IUCN Threatened Species.

Uses

This book has many examples of the applications of MRIO but here we briefly highlight three applications of Eora: carbon footprints, biodiversity fingerprints, and supply chain analysis of conflict minerals.

Carbon footprints are a headline use of MRIO. As mentioned, China has proposed that responsibility for territorial GHG emissions should be adjusted for net exports. Realizing this in policy will require the use of robust, defensible IO tables. Eora's high sector detail, or resolution, and associated confidence information address this need. High sector resolution is preferable (though admittedly provides decreasing marginal benefit) because with low sector resolution, products similar in their economic use, but different in terms of their carbon intensity, get lumped together. For example consider Australia. It exports both aluminium (a carbon-intensive product) and copper (much less carbon intensive). Assume for simplicity that all the aluminium goes to Japan and all the copper goes to France. If aluminium and copper were lumped together in a single Nonferrous Metals sector then it would be impossible to distinguish the carbon footprint of aluminium sent to Japan from the smaller carbon footprint of copper sent to France. High sector detail improves accuracy. Uncertainty information further helps analysts know the degree of confidence they can place in the MRIO table (this is discussed below). Providing accurate carbon footprints was a major motivation for undertaking this project.

A recent innovative use of Eora has been to trace products implicated in biodiversity loss to final consumers. Fishing, forestry, agriculture, and pollution in tropical developing countries are heavily responsible for biodiversity threats. Using data on species threats from the IUCN Red List of threatened and endangered species we attributed species threats to the corresponding economic sectors causing those threats, then used the MRIO to trace these implicated commodities to final consumers. The study confirmed that first-world consumer demand drives biodiversity threats in developing countries. To date tracing implicated supply chains has been labour intensive. Certification bodies providing sustainability labels must manually trace every supply chain. This approach is highly accurate but labour intensive, and because it is not systematic some impacting processes or parallel supply chains could be easily overlooked. MRIO-based supply chain analysis can significantly help improve consumer sustainability branding.

An example of supply chain analysis using MRIO is a study on columbite-tantalite (coltan) production from the Democratic Republic of Congo. Tantalum is a rare-earth metal prized for use in high-energy lightweight capacitors such as those found in mobile phones. In 2000, demand for tantalum grew sharply. Artisanal miners working surface deposits in Congo could feed this demand, but the resulting influx of cash inflamed the existing civil war and profits from the industry directly fuelled and fed violent conflict. On the ground reports documented a trade route taking coltan to neighbouring Rwanda where it was shipped onward to a processing plant in Kazakhstan (one of the few buying on the spot market). Using the Eora MRIO we have traced the supply chain onward from there to find out which companies and consumer products likely ended up using this conflict mineral. Reconstructing events over a decade past is admittedly an academic exercise, but as global MRIOs become easier to build, more accurate, and more current, we can expect detailed supply chain analysis such as this to be only a click away.

Development Using AISHA: A Tool to Build MRIOs

To build Eora we wrote AISHA (An Integrated System for building Harmonized Accounts), a software package that automates the construction of an MRIO. AISHA provides a data processing language to help connect together disparate IO data into a single table. It also provides an optimization module to reconcile conflicting input data and balance the global IO table.

Mathematically the Eora MRIO table is a multi-dimensional matrix, or tensor: the basic two-dimensional IO table is stacked vertically for each of five price layers (allowing prices to be differentiated between the basic labour and material price, taxes, subsidies, transport and middleman costs, and consumer prices). This 3-dimensional stack is available over many years, so time is the 4th dimension. On each of the IO table sheets we use a tree structure to locate countries, blocks (transaction matrix, final demand block, etc.) and sectors within each block. This tree structure helps us paste together national IO tables with heterogeneous structures (industry-industry, commodity-commodity, and supply-use tables) and heterogeneous classification schemes.

The AISHA data processing language exposes this tree structure to allow us to easily address specific areas of this tensor (e.g. the domestic IO table or the trade block containing the bilateral sector-wise trade between two countries). Tightly integrated with the language is a deep library of correspondence matrices that help aggregate, disaggregate, and translate data from one classification into another.

Correspondence matrices are an indispensible tool for joining together disparate data by translating between classifications. For example, exports from Algeria are reported in a 25 sector classification but imports into France are reported in a 61 sector classification. Correspondence matrices allow us to connect these. A correspondence matrix consists of an input vector (along the rows) and a target vector (along the columns). Along each row are percentages distributing the value from one element in the source vector among one or more elements in the target vector. In order to avoid the need to create N^2 correspondence matrices to link together each of N classifications, we used a classification ('X') with the highest possible level of detail and built correspondences with that. Thus instead of needing to build every correspondence matrix between every combination of country classification N_a and N_b we could go from $N_a \Rightarrow X$ then $X \Rightarrow N_b$ using just N correspondence matrices.

For each raw dataset input we have written a script to instruct AISHA where to place the data in the MRIO tensor and what correspondence matrices to apply. AISHA executes these scripts and sews the patches of raw data together into a single large global table.

This initial table is not yet final because most often it is not well balanced or information about certain data points may be conflicting. Most commonly, the reported sum of exports from country S rarely equals the total amount that all other countries report receiving from S. An imbalanced table implies that flows are entering from, or exiting to, an unknown source. Often this disagreement comes from the original raw data: for example country R says it exports $100 to country S, but country S reports importing just $90 from R.

To handle conflicting and disagreeing data we treat each data point not as a fixed value but as an estimate with an associated standard deviation. This method is well studied in the IO literature and is essentially justified because most statistical bureaux are themselves not 100% confident of the accuracy of every data point in their national IO tables. Given then that values have an associated uncertainty we can use optimization software to find a balanced MRIO that minimally disturbs the initial table.

To estimate the standard deviation of each value we use a number of heuristics. Based on experience we know that the larger, more important transaction values are typically better-measured and thus more reliable than small values. By default AISHA assigns a relatively small standard deviation to large data points and a larger standard deviation to smaller values (this range is set by the user). The AISHA user may also specify confidence. In the case of Eora we assign a higher reliability to national statistical agency data than to UN data, and higher reliability to UN economic data than to trade data. Part of the AISHA output is a series of reports detailing the effects of the optimizer and the degree to which various input data disagreed or had to be adjusted in order to achieve a balanced table. We use these reports to identify hotspots where we need to find better quality data.

Since no commercial optimization software was able to solve a system as large as the Eora MRIO, we had to develop our own optimization software. The Eora matrix itself has ~1.1 billion elements (9 GB) and it is subject to approximately 70 million constraints (balancing constraints, conflicting data, and non-negativity constraints). In total the optimizer must consider over 40GB of data. Building and balancing the Eora MRIO is computationally intensive: it requires approximately 2 weeks to rebuild Eora on a purpose-built cluster with 66 3GHz cores, 600GB of RAM and 15Tb of mixed SSD/RAID storage. The table is rebuilt and balanced each time we adjust or add input data.

Next Steps

The Eora MRIO marks a major step forward in the field of MRIO construction. It provides an unprecedented level of detail and coverage. The novel use of a data processing language and optimizer allow us to incorporate national data in original detail and still end up with a balanced global IO table. We will continue to incorporate new raw data into Eora in order to improve the results and keep the MRIO current. Leontief's original IO table of the US economy took him and 400 staff two years to construct. Most national IO tables built today still require substantial manual effort. Using modern hardware and software we have been able to bind together national and UN data into a global MRIO for a fraction of that cost.

We anticipate MRIO analysis will take a central role in environmental policymaking. From calculating trade-adjusted carbon footprints to conducting supply-chain analysis, high-resolution MRIO tables can help us better understand our globalized economy.

The Eora MRIO is free for non-commercial use and is available online at http://worldmrio.com

Further reading

Lenzen, M., Wood, R., & Wiedmann, T. (2010). Uncertainty analysis for Multi-Region Input-Output models - a case study of the UK's carbon footprint. *Economic Systems Research , 22*, 43-63.

Leontief, W. (1974). Structure of the World Economy. *American Economic Review , LXIV* (6), 823-834.

Su, B., Huang, H. C., Ang, B. W., & Zhou, P. (2010). Input-output analysis of CO2 emissions embodied in trade: The effects of sector aggregation. *Energy Economics , 32* (1), 166-175.

Acknowledgments

The author would like to thank and acknowledge his colleagues on the Eora project, Arne Geschke and Keiichiro Kanemoto and the Chief Investigator Prof. Manfred Lenzen. The writing of this chapter was supported by the Australian Research Council through Discovery Projects DP0985522 and DP130101293.

Chapter 8: Simplification of Multi-Regional Input-Output Structure with a Global System Boundary: Global Link Input-Output Model (GLIO)

Keisuke Nansai, Shigemi Kagawa, Yasushi Kondo, and Sangwon Suh

MRIO for Highly Accurate Global Environmental Input-Output (EIO) Analysis

In this paper we examine Multiregional Input-Output (MRIO) models, and demonstrate how they can be applied to environmental input-output (EIO) analyses of the global supply chain. Examining the global supply chain requires careful consideration of the economic activities of the countries of the world. The world economy produces innumerable goods and services. The raw materials and processes that are used to produce these goods and services vary according to geographical location. Therefore, in order to conduct accurate analyses, precise classifications of the raw and fabricated materials that are used to produce these goods and services are required. In addition, environmental analysis typically requires data on the environmental burden associated with the production of those goods and services. Consequently, numerous types of environmental burden data need to be considered in comprehensive environmental impact assessments. These environmental data might include data on energy consumption, greenhouse gases (GHG), air and water quality, soil contamination, municipal and industrial wastes, and the exploitation of water and natural resources. In addition, annual changes in the economic activities of individual countries and international trade mean that conducting analyses based on the latest global supply chain data, as well as monitoring the dynamics of the global economy and environmental burdens, requires the continuous development of MRIO models.

In general, the development of such ultra-high-resolution MRIO models can be costly[1]. It is therefore important to investigate ways in which MRIO models can be simplified. For example, suppose that rather than developing an MRIO time-series, we create an MRIO for a single year and analyze only one type of environmental burden. If we restrict the target years and environmental burdens to suit the purposes of a given analysis, then we can reduce the amount of data that needs to be prepared, and consequently, the costs associated with MRIO development. So what can be done to further reduce the amount of data that needs to be prepared? One possibility is to simplify the MRIO table structure. This chapter outlines typical MRIO simplification methods and their characteristics, and introduces the structure of the global link input-output (GLIO) model, which is based on the authors' simplified MRIO model.

Approaches to Simplifying MRIO Models

The MRIO model for global environmental analysis is based on a table that includes all the countries and regions in the world with detailed definitions of all types of goods and commodities. As an example, let us consider Table 1, which is a simplified representation of an environmental MRIO table. In Table 1, for simplification, final demand categories (e.g., household consumption, governmental consumption, etc.) and value-added categories (e.g., labour, income, profit, etc.) are omitted from the table, which shows a world comprised of three countries, A, B, and C. In this example, global economic transactions involve only two commodities: goods and services. The values shown indicate the annual transaction value ($/year), and the values in each column show the purchases of each commodity that are required by each country to produce its commodities. The table also provides information on the environmental burden (in this table, GHG emissions) associated with the production of each commodity.

Table 1: Standard MRIO structure for three countries producing two commodities.

	[$]	A		B		C	
		goods	services	goods	services	goods	services
A	goods	1	2	3	4	5	6
	services	7	8	9	10	11	12
B	goods	13	14	15	16	17	18
	services	19	20	21	22	23	24
C	goods	25	26	27	28	29	30
	services	31	32	33	34	35	36
Env	GHG (t)	37	38	39	40	41	42

---

[1] An exception to this is the Eora database discussed in Chapter 7

For example, the third column in Table 1 shows which commodities are necessary to produce a good in country A. To produce its own goods, country A purchases $1- and $7-worth of country A's goods and services, $13 and $19-worth of country B's goods and services, and $25 and $31-worth of country C's goods and services, respectively. The total purchases by country A amount to $96. The GHG emissions resulting from the production of country A's goods is 37 t. Note that the total value of the global transactions shown in Table 1 is $ 666 of the accumulated amount from $1 to $36 and the total emissions for all countries is 237 t of the accumulated total from 37 t to 42 t.

Table 1 clearly shows that the commodities required to produce goods and services differ depending on the country (A, B, and C). In order to construct Table 1, we need the survey data for the 7×6=42 cells in the table. Alternately, we would need estimates for these values obtained by other methods, but this would entail an increase in table construction costs. There are three main approaches to reducing these costs, each of which is explained briefly below to highlight the relative advantages and disadvantages.

1. Reducing the resolution of commodity classification

The first method for reducing table construction costs is to reduce the number of sectors by aggregating detailed sector classifications and reducing the sector resolution. For example, Table 2 shows the aggregated values for goods and services for each country that was listed separately in Table 1 by defining the commodity sum as the total of goods + services. The number of data cells that needed to be estimated to construct this table was 4×3=12, which is equivalent to 28.5% of the 42 data cells that were required to construct Table 1 (12/42×100 = 28.5%).

In Table 2, the commodities required to calculate the sum of the annual production in country A are given by the sum of $18 for country A (1+2+7+8), the sum of $66 for country B (13+14+19+20), and the sum of $114 for country C (25+26+31+32). The GHG emissions based on the sum of country A is 75 t (=37+38). This sector aggregation implies that countries A, B, and C produce composite commodities that don't distinguish between goods and services, although the global transactions and the global emissions will still be $ 666 (= 18+26+34+66+74+82+114+122+130) and 237 t (= 75+79+83) respectively. Since countries A, B, and C are clearly differentiated, they can be subjected to analyses that focus on national differences, such as country-specific carbon footprint analyses. On the other hand, because no distinction is made between goods and services, this table data would not be suitable for analyses that focus on differences in production technologies, such as product life cycle assessments (LCA).

Table 2: Simplification of MRIO table by reducing sector resolution

	[$]	**A** Composite commodity	**B** Composite commodity	**C** Composite commodity
A	Composite commodity	18	26	34
B	Composite commodity	66	74	82
C	Composite commodity	114	122	130
Env	GHG (t)	75	79	83

2. Reducing the resolution of country distinctions

The second method for reducing table construction costs is to reduce the number of countries by aggregating multiple countries and reducing country resolution. For example, Table 3 combines countries B and C, which were shown separately in Table 1, to create a composite country (B&C). This results in 5×4=20 data cells that need to be estimated and reduces the burden of table creation from 42 to 20 (i.e. 20/42×100= 47.6% compared to Table 1). For instance, the developments of OECD (see Chapter 9), WIOD (see Chapter 5), EXIOPOL (see Chapter 6) and GTAP (see Chapter 3) employ this method.

In Table 3, the commodities required for the annual production of goods by composite country B&C are worth $8 (3+5) of country A's goods and $20 (9+11) of its services, and worth $88 (15+17+27+29) of country B&C's goods and $112 (21+23+33+35) of its services. The GHG emissions associated with the production of composite country B&C's goods are 80 t (39+41). The total value of the transactions shown in this table is $666 (=18+26+34+66+74+82+114+122+130) and the GHG emissions are 237 t (=75+79+83), reflecting a global economy that is similar in scale to that shown in Table 1. Since the differences in the production structures for goods and services are clearly differentiated, these data can be subject to commodity comparisons and product LCA. However, because the individual emissions of countries B and C are unknown, the data are not appropriate for country-specific analyses.

Table 3: Simplification of MRIO table by reducing regional resolution

[$]		A		Composite country (B & C)	
		Goods	Services	Goods	Services
A	Goods	1	2	8	10
	Services	7	8	20	22
B & C	Goods	38	40	88	92
	Services	50	52	112	116
Env	GHG (t)	37	38	80	82

3. Reducing the number of countries considered by the model

The third method simply involves reducing the number of countries being considered without combining either the commodity classifications or the sectors of the countries. For example, Table 1 describes the production structures for countries A, B, and C, that is, the commodities required for production in each country, but Table 4 only considers countries A and B, eliminating the economic activities of country C from the global economy. This yields an input-output (IO) analysis that only includes countries A and B (creating a 4×4 matrix comprised of data for A and B), reducing the burden of table creation from 42 to 20 (i.e. 20/42×100= 47.6% compared to that in Table 1). This is an effective method for reducing costs when trade statistics for country C are not available for use in an IO table. Examples using this method are MRIOs developed by IDE-JETRO (see Chapter 16).

In Table 4, countries A and B are clearly differentiated, as are goods and services, making it possible to conduct analyses that reflect national features and characteristics of commodity production technologies. That is, these data are well suited to comparisons of the overall carbon footprints of countries A and B, and also for conducting LCA of goods in countries A and B. However, because country C is excluded, the total value of the transactions in this table is $184 ($666 in Table 1) and GHG emissions are 154 t (237 t in Table 1). Thus, in this case the analysis is based on a global economic scale that differs from the simplified MRIO models depicted in Tables 2 and 3. Consequently, if imports from country C are used extensively in the production of commodities in countries A and B, or if the global proportion of the GHG emissions from country C is high, then the results obtained using this table may understate the carbon footprint of country A or B, or the LCA values of their products.

Table 4: Simplification of MRIO table by reducing regional system boundaries

	[$]	A Goods	A Services	B Goods	B Services	C
A	Goods	1	2	3	4	
A	Services	7	8	9	10	
B	Goods	13	14	15	16	
B	Services	19	20	21	22	
C						
Env	GHG (t)	37	38	39	40	

Global Link Input-Output Model (GLIO)

Let us consider an example of a simplified environmental MRIO model, referred to hereafter as a GLIO model. The GLIO model was developed to analyse environmental burdens that are generated within global supply chains associated with Japanese goods and services. It is well known that among developed countries, Japan is a resource-poor country. This means that the Japanese economy is highly dependent upon imports of natural resources such as crude oil, natural gas, coal and mined minerals. Furthermore, imports of both raw and fabricated materials are increasing due to increased specialization associated with the establishment of international divisions as well as the production fragmentation of Japanese companies overseas. These factors motivated us to estimate the global environmental burden of Japanese commodities that directly and indirectly account for a large proportion of imports.

The GLIO model simplifies the MRIO structure as shown in Table 1 by partly reducing the resolution of commodity classification or reducing the number of sectors in each country except for the country that is the focus (e.g. Japan). For the simplification, the GLIO model does not employ the aggregation of multiple countries into a composite country and the reduction of regional system boundaries by excluding some countries.

Table 5 illustrates a generalized GLIO framework. For this example, we will focus on country A to apply a GLIO model to an LCA of country A's goods and services. Thus, as in Table 1, we provide a detailed description of the inputs of each commodity required for the production of goods and services by country A, as well as the GHG emissions associated with that production. However, if we combine the goods and services of countries B and C, such that the sum = goods + services, then we can reduce the number of data points that need to be estimated to $7 \times 4 = 28$ (i.e. $28/42 \times 100 = 66.6\%$ compared to those in Table 1), decreasing table creation costs. The total value of the transactions shown in this table is $666 and the GHG emissions are 237 t, reflecting a global economy that is similar in scale to that shown in Table 1. However, a special method of accounting must be employed to express final demand (see Nansai *et al.*, 2012a for details).

Table 5: Simplification of MRIO table using a GLIO framework

[$]		A		B	C
		Goods	Services	Composite commodity	Composite commodity
A	Goods	1	2	7	11
	Services	7	8	19	23
B	Goods	13	14	31	35
	Services	19	20	43	47
C	Goods	25	26	55	59
	Services	31	32	67	71
Env	GHG (t)	37	38	79	83

The difference between Tables 2, 3, 4, and 5 essentially comes down to whether the transaction in goods and services is depicted as a square matrix. Since the transaction in goods and services is depicted as a square matrix in Tables 2, 3, and 4, these tables can be directly applied to the conventional environmental IO analysis that requires a square matrix of the transaction for describing a linear system of equations to ensure the existence of any supply level of each commodity for an arbitrary final demand vector (see Chapter 19 for details). Conversely, because Table 5 is rectangular, a special conversion must be employed to make it square. If the rows of goods and services for countries B and C are combined, then we can create a square as shown in Table 2, but this will make it impossible to estimate the GHG emissions in a way that precisely reflects the production structure of goods and services in country A, which is one of the aims of the analysis. Fortunately, the GLIO model can be used in such instances.

The GLIO model converts the transaction values of goods and services in the rows for countries B and C into embodied domestic GHG emissions; which are the GHG emissions emitted directly and indirectly in each country that are associated with each production of goods and services. For example, we here assume the embodied domestic GHG emissions of each commodity as follows. If $1 of country B's goods and services are produced, then:

Domestic GHG emissions embodied in producing $1 of goods in country B will be 0.5 t (0.5t/$).

Domestic GHG emissions embodied in producing $1 of services in country B will be 0.4 t (0.4t/$).

If $1 of country C's goods and services are produced, then:

Domestic GHG emissions embodied in producing $1 of goods in country C will be 0.3 t (0.3t/$).

Domestic GHG emissions embodied in producing $1 of services in country C will be 0.2 t (0.2t/$).

The value of country B's goods required to produce the goods of country A is \$13, but if this is expressed as the embodied domestic GHG emissions, then the result is 13×0.5=6.5 t. Similarly, if \$19 is input into country B's services, then the conversion is 19×0.4=7.6 t of GHG. If the same conversion is performed for country C, then we can create hybrid Table 6, which expresses country A's row transactions in dollars, and country B and C's row transactions in tons. In Table 6, the unit for expressing the volume of transactions needed changes, but the table shows the production structure of goods and services of country A at the same level as Table 1. Furthermore, the differences in GHG emissions between goods and services in countries B and C are also reflected. Country B and country B's inputs, and country C and country C's inputs correspond to intermediate inputs, and when they are converted to embodied domestic GHG emissions, the result is zero (see Nansai *et al.* 2012a for details).

Although the embodied domestic GHG emissions (0.5 t/\$, 0.4 t/\$, 0.3 t/\$, and 0.2 t/\$) used above are just hypothetical values, in practice, those values or the domestic GHG emissions embodied in producing \$1 of goods (or services) in country B (or C) can be set by a conventional EIO analysis with domestic system boundary of country B (or C). They can be also substituted by carbon footprint data or life cycle inventory data for life cycle assessment of country B (or C), that show total GHG emissions in producing \$1 of goods in country B (or C).

Table 6: Rectangular GLIO table framework

		A		B	C
	[\$]	Goods	Services	Composite commodity	Composite commodity
A	Goods	1	2	7	11
	Services	7	8	19	23
B	Goods	6.5 t	7 t	0 t	17.5 t
	Services	7.6 t	8 t	0 t	18.8 t
C	Goods	7.5 t	7.8 t	16.5 t	0 t
	Services	6.2 t	6.4 t	13.4 t	0 t
Env	GHG (t)	37	38	79	83

Finally, by aggregating the rows of goods and services for countries B and C, as shown in Table 6, we can construct the square matrix shown in Table 7. In this table, the unit of outputs from country A is monetary and the unit of outputs from countries B and C is mass of GHG emissions. Such an input-output model consisting of flows with different units is called a mixed-units model and has a long tradition in an input-output analysis. Various analytical methods used in conventional EIO analyses can also be applied to the GLIO table shown in Table 7 (see Nansai *et al.* 2009 and 2012a for details). The relationship between country B's total emissions of 79 t and country B's exported emissions can be characterized as follows. Country B has total emissions of 79 t (see the last row in column B of Table 7). It emits 29.1 t (=14.1+15) for its exports to country A, and 36.3 t for its exports to country C. The remaining 13.6 t (79.0-29.1-36.3) can be attributed to the domestic final demand of country B. Country C's emissions can be clarified in a similar way. Importantly, unlike Table 4, an EIO analysis based on Table 7 reveals that overall global GHG emissions are maintained.

Table 7: Square GLIO table framework

[$]			A		B	C
			Goods	Services	Composite commodity	Composite commodity
A		Goods	1	2	7	11
		Services	7	8	19	23
B		Composite commodity (t)	14.1	15	0	36.3
C		Composite commodity (t)	13.7	14.2	29.9	0
Env		GHG (t)	37	38	79	83

Case Study: GLIO Model Centred Japanese Economy

The authors have developed a GLIO model that focuses on the Japanese economy. Using Table 6 as an example, if country A represents Japan and its goods and services, then the country would have 406 sectors. Countries B and C correspond to the foreign sector, consisting of 230 countries and regions having 111 goods and services that are indicated as rows of the foreign countries. Prepared environmental burden data include energy consumption, all six greenhouse gases that Japan has committed itself to reducing as part of the Kyoto Protocol (i.e. CO_2, methane (CH_4), nitrous oxide (N_2O), perfluorocarbons (PFCs), hydrofluorocarbons (HFCs), and sulphur hexafluoride (SF_6)), as well as air pollutants (nitrogen oxide (NO_x) and sulphur oxide (SO_x)). Time series data for the GLIO model have been compiled for the years 1990, 1995, 2000, and 2005. For details regarding data preparation and sample applications, see Nansai *et al.* (2012a, 2012b).

Further readings

Andrew, R., Peters, G. P., & Lennox, J. (2009). Approximation and Regional Aggregation in Multi-Regional Input-Output Analysis for National Carbon Footprint Accounting. *Economic Systems Research, 21* (3), 311-335.

Lenzen, M. (2011). Aggregation Versus Disaggregation in Input-Output Analysis of the Environment. *Economic Systems Research, 23* (1), 73-89.

Lenzen, M., Wood, R., & Wiedmann, T. (2010). Uncertainty analysis for Multi-Region Input-Output models - a case study of the UK's carbon footprint. *Economic Systems Research, 22*, 43-63.

Majeau-Bettez, G., Strømman, A. H., & Hertwich, E. G. (2011). Evaluation of Process and Input-Output-based Life Cycle Inventory Data with Regard to Truncation and Aggregation Issues. *Environmental Science and Technology, 45*, 10170-10177.

Nansai, K., Kagawa, S., Kondo, Y., Suh, S., Inaba, R., & Nakajima, K. (2009). Improving the Completeness of Product Carbon Footprints Using a Global Link Input-Output Model: The Case of Japan. *Economic Systems Research, 21* (3), 267-290.

Nansai, K., Kagawa, S., Kondo, Y., Suh, S., Nakajima, K., Inaba, R., et al. (2012a). Characterization of economic requirements for a "carbon-debt-free country". *Environmental Science & Technology, 46* (1), 155-163.

Nansai, K., Kondo, Y., Kagawa, S., Suh, S., Nakajima, K., Inaba, R., et al. (2012b). Estimates of Embodied Global Energy and Air-Emissions Intensities of Japanese Products for Building a Japanese Input-Ouput Life Cycle Assessment Database with a Global System Boundary. *Environmental Science & Technology, 46* (16), 9146-9154.

Su, B., Huang, H. C., Ang, B. W., & Zhou, P. (2010). Input-output analysis of CO2 emissions embodied in trade: The effects of sector aggregation. *Energy Economics, 32* (1), 166-175.

Acknowledgement

This research was supported in part by Environment Research & Technology Development Fund (S-6-4) of the Ministry of Environment, Japan

Chapter 9: The Global Resource Accounting Model (GRAM)

Kirsten S. Wiebe, Christian Lutz, Martin Bruckner, and Stefan Giljum

Introduction

Increasing global trade and, as a result, increasingly internationalized production chains lead to a separation of the locations of production and consumption of goods and also services. While in 2005 in the OECD countries only about 40% of global energy-related carbon emissions were generated during production processes, the goods consumed within the OECD countries embodied more than 60% of global carbon emissions. This similarly holds for natural resources: about one third of the material and energy resources used by the OECD are imported. These numbers show that a large part of the environmental burden of resource extraction for and generation of greenhouse gas (GHG) emissions during the production of goods and services that are consumed in the industrialized countries falls onto the newly emerging and developing economies. In turn, this also means that the OECD countries become increasingly dependent on production in the other countries. Using multi-regional input-output models, such as the Global Resource Accounting Model GRAM, which is presented in this chapter, it is possible to calculate the embodied material, energy and pollution flows of international trade and also consumer and producer responsibilities.

The Global Resource Accounting Model

The Global Resource Accounting Model (GRAM) has been developed in the course of several projects. The main objective of the first version of GRAM was to estimate indirect material flows of traded products (measured as their raw material equivalent) making it possible to calculate and analyse material flow-based indicators in a global perspective, considering comprehensive material balances on the national level, which take into account all up-stream material requirements of imports and exports. The model was composed of about 50 single-country input-output models linked through bilateral trade. This linking of the individual country models made it impossible to get a solution for all countries simultaneously, so that the final result was only an approximation of the actual solution. Global material exports did not equal global material imports, but were sufficiently close.

For the second and most recent version of GRAM, the set-up of the model and the solution procedure changed significantly, though it is still based on the same data. The input-output data and bilateral trade data are now used to construct a multi-regional technical coefficient matrix and the corresponding final demand matrix. The system is then solved at once by explicitly applying the Leontief inverse[1], which is state-of-the art in the current MRIO literature to take all global production chains into account. This solution procedure ensures that global emissions embodied in exports equal global emissions embodied in imports. This version of GRAM was first applied to calculate the carbon emissions embodied in Austria's international trade. Since then GRAM has been applied to the various issues that are described below and documented in the literature mentioned in the section on further reading.

The Structure of GRAM

GRAM is a multi-regional input-output model covering 53 countries and two regions (OPEC and the rest of the world – RoW) and 48 sectors per country/region. It is a MRIO defined in the narrow sense in Chapter 1, because the trade blocks of the global matrix are estimated using bilateral trade data. The country coverage is displayed in Figure 1. Including the region Rest of the World ensures global closure of the model. The model is based on the OECD structural analysis (STAN) input-output and bilateral trade data. The input-output data distinguish between 26 industrial sectors and 22 service sectors, while the bilateral trade data are only available for 25 industrial sectors (they do not differentiate between energy and non-energy mining and quarrying) and one aggregate service sector.

---

[1] This is a rather important, but technical detail.

Figure 1: Country coverage of GRAM

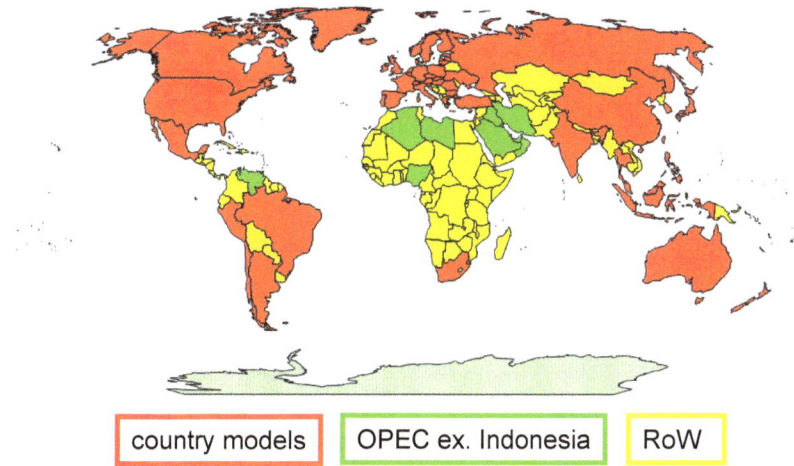

| country models | OPEC ex. Indonesia | RoW |

GRAM does not contain a multi-regional input-output table, hence, it does not directly display the absolute intra- and international monetary flows between the industries; rather, it *models* intermediate input shares. Intermediate input shares are the fractions each sector provides to the total inputs of the sector in question. The intermediate input shares are constructed from national input shares and import shares as will be explained below. The difference to most other constructed MRIO models is the assumption in GRAM that the intermediate input structure given by these intermediate input shares is correct. This however implies that output and also value added, which are calculated within the model, may not perfectly fit the output and value added data given in the OECD statistics. Most other models in contrast assume that both final demand and output data as well as value added are correct and use balancing procedures to alter the intermediate input share matrix, to fit these constraints. The construction of the global input coefficient matrix from the original OECD input-output tables is explained in more detail now:

The OECD input-output tables (IOTs) consist of five matrices:

- 2 'industry × industry' intermediate demand matrices, one displaying flows between domestic industries and one displaying the monetary value of deliveries from foreign industries to domestic industries.

- 2 'industry × consumer' group final demand matrices, again one with flows from domestic industries to domestic consumers (private households, government, investment and exports) and one with flows from foreign industries to domestic consumers.

- 1 primary inputs matrix that includes taxes and subsidies, compensation of employees, depreciation, profits and total output for each industry.

Given these data it is possible to calculate a domestic input coefficient matrix and an imported input coefficient matrix for each country. An input coefficient matrix includes in each row the shares of each sector's inputs into the sector corresponding to the column. The sum of all input shares in each column plus the share of value added should add up to one. The matrices on the left in

Figure 2 show the domestic (top) and import (bottom) input coefficient matrix of country A and two sectors G1 and G2. Firms in sector G1 receive on average half of their inputs (0.5) from themselves and other domestic firms in sector G1, and 10% (0.1) of their inputs from domestic firms in sector G2. They import 20% of their inputs from firms in sector G1 and 10% of their imports from firms in sector G2.

Figure 2: Example for an import coefficient matrix for country A

domestic

A	G1	G2
G1	**0.5**	**0.2**
G2	**0.1**	**0.2**

imported

A	G1	G2
G1	**0.2**	**0.1**
G2	**0.1**	**0.3**

Domestic and import coefficient matrix of country A

	A	G1	G2
B	G1	**0.10**	**0.05**
	G2	**0.04**	**0.12**
C	G1	**0.08**	**0.04**
	G2	**0.03**	**0.09**
D	G1	**0.02**	**0.01**
	G2	**0.03**	**0.09**

Import matrix split up among other countries

Figure 3: Example for an import share matrix of goods G1 and G2

exporting country / importing country

G1	A	B	C	D
A	-	0.2	0.3	0.2
B	0.5	-	0.4	0.4
C	0.4	0.5	-	0.4
D	0.1	0.3	0.3	-
total	**1.0**	**1.0**	**1.0**	**1.0**

exporting country / importing country

G2	A	B	C	D
A	-	0.1	0.2	0.3
B	0.4	-	0.4	0.3
C	0.3	0.4	-	0.4
D	0.3	0.5	0.4	-
total	**1.0**	**1.0**	**1.0**	**1.0**

To construct the multi-regional input coefficient matrix, the import coefficient matrix of each country, say country A, has to be split according to the share each of the other countries has in the imports of country A. Figures 2 and 3 illustrate this procedure. First, the import share matrices – as displayed in Figure 3 for a four-country (A, B, C, D) two-sector (G1 and G2) example – were constructed from the bilateral trade data. Here, country A imports 50% (0.5) of goods produced by industry G1 from country B, 40% (0.4) from country C and 10% (0.1) from country D, as displayed in the first column of the G1 matrix in Figure 3.

Figure 2 shows on the left the domestic and import input coefficient matrix of country A. The import coefficient matrix is now split among the other countries using the import share matrices from Figure 3. The result is displayed on the right in Figure 2. The share of 0.1 of imported goods from sector G2 as inputs to sector G1 (dotted fields in the matrix on the left/bottom of Figure 2) is split across countries B, C and D using the first column of the import share matrix of G2 in Figure 3, that is with shares 0.4, 0.3 and 0.3, respectively, resulting in the input shares 0.04, 0.03 and 0.03 (dotted) in the matrix on the right in Figure 2. This procedure is also applied to the other entries of the import input share matrix.

The final demand matrices are split accordingly, after converting these from national currency to USD using market exchange rates from the International Financial Statistics of the International Monetary Fund. For each country, only household final consumption, government final consumption, gross fixed capital formation and changes in inventories are considered as final demand categories. The sectoral final demand exports are determined by all other countries' imports.

The data directly included from the OECD STAN database therefore are the final demand matrices excluding sectoral exports. The OECD input-output tables and bilateral trade data are used to estimate the global technical coefficient matrix. Sectoral output, and with it sectoral value added, is calculated within the model. The two basic assumptions underlying this modelling approach are, first, that the economic structure is best represented using the unbalanced coefficients and, second, that the reported import data are superior to the reported export data. These assumptions imply, first, that we did not use a balancing procedure such as a RAS method, and second, that the other countries' imports determine a country's export. The model implicitly considers transit trade, which is reported in the countries' import final demand tables.

The difference to the statistically reported sectoral production is within ±10% for half of the sectors[2], for which data are available in the OECD STAN database.

Applications of GRAM

GRAM has been used for historical calculations of emissions and material embodied in international trade and in combination with the energy-environment-economy model GINFORS (Global INterindustry FORecasting System). The results can also be used to allocate the environmental burden from producers to consumers of goods and services.

---

[2] See Wiebe et al. (2012a) in section *Further Readings*.

Consumption-based carbon accounting

GRAM, just like several other MRIOs, has been used for the calculation of consumption-based carbon emissions. The emissions data enter the model in form of sectoral emission intensities **E**. That is, for each sector in each country we calculate the ratio of carbon emissions per unit of sectoral output using CO_2 emissions data provided by the International Energy Agency (IEA) and sectoral output data that were calculated in the model. These emissions intensity coefficients are then used in the model to calculate the pollution embodied in international trade, and thus, the pollution associated with each country's final demand.

The pollution matrix **P** is displayed in Figure 6 for a four-country two-sector example. $p_{AA(G1)} + p_{AA(G2)}$ for example includes the emissions of Austria's (country A) industry and service sectors that occur while Austria's industry is producing goods that are consumed within Austria. The remaining numbers in the first two rows of the pollution matrix, highlighted in light grey and grey in Figure 4, contain emissions that occur while Austria's industry produces goods for final consumption in the other countries. The sum of the light grey/grey entries in these two rows is the emissions associated with Austria's exports. Similarly, the sum over all sectors and countries in the first column, highlighted in dark grey, excluding the domestically produced emissions in the dotted part, are the emissions embodied in Austrian imported final consumption goods. Selected results for emissions embodied in Austria's trade are presented in Chapter 20.

From this matrix it also becomes clear how the embodiment of emissions in final goods and services can be calculated: The total emissions that occur during production in Austria are the sum of the dotted and light grey/grey parts of the table. The emissions embodied in the goods and services consumed in Austria are the sum of the dotted and dark grey parts of the table.

Figure 4: Pollution matrix P

			importing country				Emissions embodied in exports of
			A	B	C	D	
exporting country & sector	A	G1	PAA(G1)	PAB(G1)	PAC(G1)	PAD(G1)	country A industry G1 PAB(G1)+PAC(G1)+PAD(G1)
		G2	PAA(G2)	PAB(G2)	PAC(G2)	PAD(G2)	country A industry G2 PAB(G2)+PAC(G2)+PAD(G2)
	B	G1	PBA(G1)	PBB(G1)	PBC(G1)	PBD(G1)	
		G2	PBA(G2)	PBB(G2)	PBC(G2)	PBD(G2)	
	C	G1	PCA(G1)	PCB(G1)	PCC(G1)	PCD(G1)	
		G2	PCA(G2)	PCB(G2)	PCC(G2)	PCD(G2)	Emissions embodied in goods produced and consumed in country A
	D	G1	PDA(G1)	PDB(G1)	PAC(G1)	PAD(G1)	
		G2	PDA(G2)	PDB(G2)	PDC(G2)	PDD(G2)	PAA(G1)+PAA(G2)

Emissions embodied in imports of
country A PBA(G1)+PBA(G2)+PCA(G1)+PCA(G2)+PDA(G1)+PDA(G2)

Material rucksacks on international trade

GRAM has also been applied to calculate material rucksacks on international trade for four material categories: biomass, fossil fuels, metals and industrial minerals, and construction minerals. Results of this exercise are also presented in Chapter 20. Calculating materials embodied in production and consumption, imports and exports is almost equivalent to the calculation of embodied carbon emissions. In general, the data included in the intensity coefficients **E** can be interchanged to include any greenhouse gas emissions that occur during production or materials used for production.

Combining GRAM and an energy-environment-economy projection and simulation model

Recently, GRAM has been applied to analyse the effect of a global climate policy scenario on emissions embodied in consumption. For that, it uses the results of a scenario analysis with the Global INterindustry FORcasting System GINFORS regarding mitigation efforts in line with the Copenhagen pledges for 2020. GINFORS is a dynamic macro-econometric input-output model that has been widely applied in analyzing environmental policy measures [3]. GINFORS projections provide GDP development, energy balances and energy-related carbon emissions for 53 countries and two regions. Sectoral production structures and trade data are available for all OECD countries, their major trading partners and the large emerging economies. The results of the scenario analysis in GINFORS are used to project the MRIO coefficient matrix and the corresponding final demand matrix as well as the energy-intensity coefficient vector of the MRIO model GRAM, which is then used to calculate consumption-based carbon emissions of the given Post-Kyoto regime until 2020.

Production-based carbon emissions will further increase in emerging economies, whereas OECD countries will have to reduce emissions according to their Copenhagen pledges. The results show that a consumption-based accounting of carbon emissions allocates more emissions to the industrialized countries than production-based accounting. However, the increasing final demand in the emerging economies may reduce net-exports and hence also the relative net-emissions embodied in these exports. Global responsibility for the lager part of carbon emissions will remain with the highly industrialized OECD countries. Still, the shift towards a higher responsibility of emerging economies is inevitable as these are growing faster and continue producing with more carbon-intensive technologies.

Other possible applications

GRAM can also be used for an in depth investigation on the relationship between trade and labor demand on occupational and qualification levels. The intensity

---

[3] See Lutz et al. (2010) in section *Further Readings*.

coefficient matrix then consists of labor intensity coefficients that include either occupations or qualifications. This adds a more socio-economic application possibility to multi-regional input-output analysis, since from qualification levels and occupations it is possible to calculate labor costs and wages and, hence, also income. It is further easily possible to extend the model to do similar calculations for different kinds of ecological footprints, such as water footprints, energy footprints and other greenhouse gas footprints.

GRAM: prospects

GRAM is based on official statistics provided by the OECD, IEA and International Monetary Fund. These international organizations will continue to make the data series available in the future, so that a continuous updating of the historical GRAM database to include the most recent data is possible. Recent joint activities of the OECD and the World Trade Organisation on the issue of value added embodied in trade and projects by the European Commission on ecological footprints show the growing interest in new applications of MRIO models like GRAM.

Further reading

Bruckner, M., Giljum, S., Lutz, C., & Wiebe, K. (2012). Materials embodied in international trade – Global material extraction and consumption between 1995 and 2005. *Global Environmental Change, 22* (3), 568-576.

Giljum, S., Lutz, C., Jungnitz, A., Bruckner, M., & Hinterberger, F. (2011). European Resource Use and Resource Productivity in a Global Context. In P. Ekins, S. Speck, P. Ekins, & S. Speck (Eds.), *Environmental Tax Reform (ETR) - A Policy for Green Growth* (pp. 27-45). New York: Oxford University Press.

Lutz, C., Meyer, B., & Wolter, M. I. (2010). The Global Multisector/Multicountry 3E-Model GINFORS. A Description of the Model and a Baseline Forecast for Global Energy Demand and CO2 Emissions. *International Journal of Global Environmental Issues, 10* (1-2), 25-45.

Nakano, S., Okamura, A., Sakurai, N., Suzuki, M., Tojo, Y., & Yamano, N. (2009). *The Measurement of CO2 Embodiments in International Trade: Evidence from the Harmonized Input-Output and Bilateral Trade Database. STI Working Paper, 2009/3 (DSTI/DOC(2009)3).* OECD. Paris: OECD.

Wiebe, K., Lutz, C., Bruckner, M., & Giljum, S. (2012a). Calculating energy-related CO2 emissions embodied in international trade using a global input-output model. *Economic Systems Research, 24* (2), 113-139.

Wiebe, K., Lutz, C., Bruckner, M., Giljum, S., & Poliz, C. (2012). Carbon and Materials Embodied in the International Trade of Emerging Economies:

A Multi-regional Input-Output Assessment of Trends between 1995 and 2005. *Journal of Industrial Ecology 16*(4), 636-646.

Resources

IEA Energy Balances of OECD and Non-OECD Countries. International Energy Agency, Paris, France.

IEA CO2 Emissions from Fuel Combustion. International Energy Agency, Paris, France.

IMF International Financial Statistics. International Monetary Fund, Washington DC, US.

OECD STAN Bilateral Trade Database. Organisation for Economic Co-operation and Development, Paris.

OECD STAN Input-Output Tables. Organisation for Economic Co-operation and Development, Paris.

Acknowledgements

The results presented in this chapter were sourced from two projects funded by the Austrian Climate and Energy Fund and the Anglo-German Foundation.

Part III: Special MRIO Frameworks

Bushfire burning
Oil on canvas 108x132cm
Dagmar Hoffmann

Chapter 10: Distribution of CO_2 Emissions in China's Supply Chains: A Sub-national MRIO Analysis

Kuishuang Feng and Klaus Hubacek

Introduction

As the world's biggest emitter, China is facing the challenge of balancing rapid growth of the economy and environmental sustainability. Rather than a homogenous country that can be analyzed at the national level, China is a vast country with substantial regional differences in physical geography, regional economic development, demographics, infrastructure, and household consumption patterns[1]. Particularly, there are pronounced differences between the well-developed coastal regions and the less developed central and western regions, which lead to large regional discrepancies in CO_2 emissions[2]. By using environmentally extended MRIO analysis, in this study we present consumption-based CO_2 emissions based on intra- and inter-regional supply chains in China. We include 26 provinces and 4 municipalities, and then aggregate them to 8 regions for ease of presentation. The 8 regions are North-East, Beijing-Tianjin, North, Central-Coast, South-Coast, Central, North-West and South-West.

---

[1] Feng, K., Siu, Y.L., Guan, D., Hubacek, K., 2012. Analyzing Drivers of Regional Carbon Dioxide Emissions for China. Journal of Industrial Ecology In press.

[2] Feng, K., Hubacek, K., Guan, D., 2009. Lifestyles, technology and CO_2 emissions in China: A regional comparative analysis. Ecological Economics 69, 145-154.

Results and discussion

Embodied CO₂ emissions in inter-regional trade

Fig. 1 shows the Central-Coast region with the largest embodied CO_2 emissions in its domestic imports, mainly from Central (148 MMT or 33% of the total embodied emissions in Central region's exports), North (127 MMT or 29% of the total) and North-West (68 MMT or 21% of the total). At the same time Central-Coast region exports only small amounts of embodied emissions to other regions. It indicates that the consumption of goods in Central-Coast is significantly relying on production in other regions and thus imposing emissions to other regions through inter-regional supply chains. Central region is ranked as the second largest importer of embodied CO_2 emissions in inter-regional trade. Its import and associated embodied emissions are from the North, Central-Coast and North-West. However, embodied emissions in imports for the Central are much smaller than the emissions embodied in its export. The North is also a net exporter of embodied emissions, exporting fairly large amounts of embodied emissions to Beijing-Tianjin (71 MMT or 16% of the total embodied emissions in the exports of the North). Beijing-Tianjin, as a political and commercial center in China, causes more than 75% of its total consumption based emissions in other regions as its local consumption highly depends on goods produced in other parts of China, while the embodied export emissions of this region are rather small. Similarly, the South-Coast is one of the wealthiest regions in China, thus imports a large amount of goods from other regions, such as the Central and South-West, and creates emissions in these regions. The North-West, particularly Inner-Mongolia, serves as China's energy hub and exports energy-intensive products, such as electricity, iron and steel and coal, to many other regions, therefore exports substantial amounts of embodied emissions. The carbon budgets for the North-East and South-West are relatively balanced with regards to other Chinese regions.

Figure 1: Embodied CO_2 emissions in inter-regional trade to satisfy regional domestic final demand (in million metric tons).

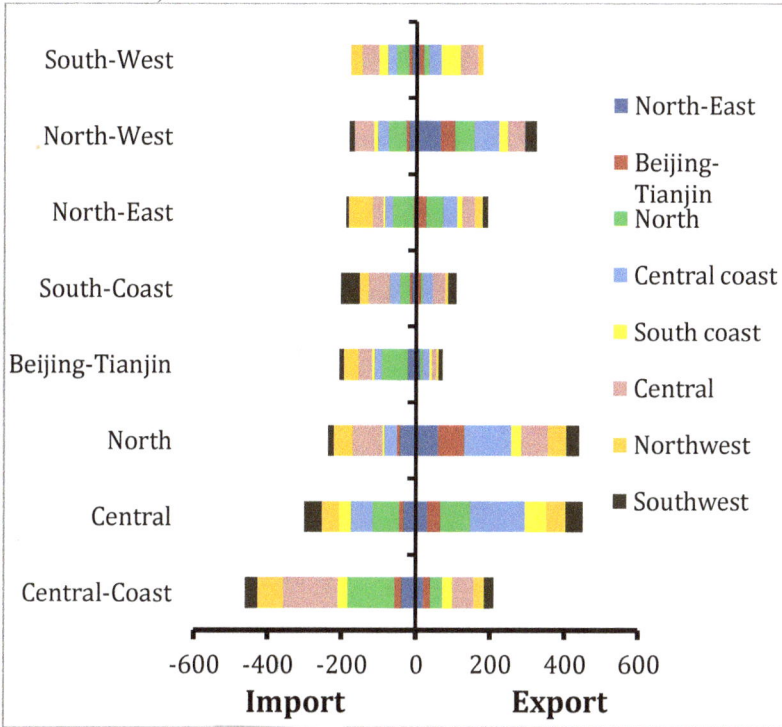

International exports are also significant drivers for China's CO_2 emissions. The wealthier Central- and South- Coast regions gain a huge amount of profits from international exports due to their advantages in terms of location and advanced infrastructure; exports contribute more than 20% to their regional GDP. However, our results show that only 60% of life-cycle emissions associated with exports occurred within their regional boarders, while the other 40% of embodied emissions were produced in other parts of China. Fig. 2 shows that international exports in Central-Coast region cause a large amount of emissions in the Central (77 MMT), North (70 MMT), Northwest (38 MMT) and North-East (17 MMT). South-Coast's international exports lead to a large amount of emissions in South-West (59 MMT), Central (45 MMT), North (22 MMT) and North-West (20 MMT). International export in North and Beijing-Tianjin also cause emissions in other regions, but the quantity is rather small. Central, North-West and South-West regions, as net embodied emissions exporter, are lack of direct connection to the international market, thus have very small emission associated with their international exports.

Figure 2: Embodied emissions in inter-regional trade driven by international export. Regions in y-axis are domestic exporters who export goods to other Chinese regions for international export.

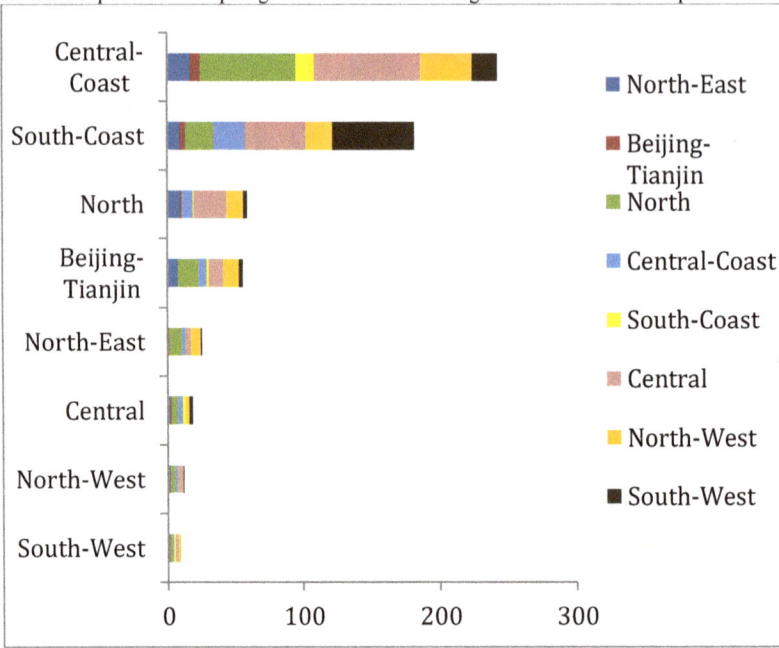

Consumption-based CO_2 emissions inventory

There are large regional disparities in terms of consumption patterns and associated per capita consumption emissions (see Figure 3). Shanghai ranked top with approximately 12 metric tons CO_2 emissions per person, which is more than six times the CO_2 emissions per person in the lowest-ranked province Guangxi. In most provinces, the construction sector is the largest contributor to total consumption emissions. This is due to the large-scale infrastructure investments across China to support (maintain) the fast growth of the economy. Building infrastructure is a carbon intensive process that requires significant inputs from other carbon intensive sectors, such as iron and steel, electricity generation and minerals sector. The contributions of the construction sector to the regional CO_2 emissions range from 14% to 51%. Figure 3 shows that in highly urbanized areas, such as Beijing, Tianjin, Shanghai, where urban population accounts for more than 80% of the total, Petroleum, chemicals and minerals (14% - 20%) and Services (10% - 24%) are also important contributors to total consumption emissions as urban households consume more services and non-basic needs products compared to rural households.

Figure 3: Per capita consumption-based CO_2 emissions by sectors (in metric tons).

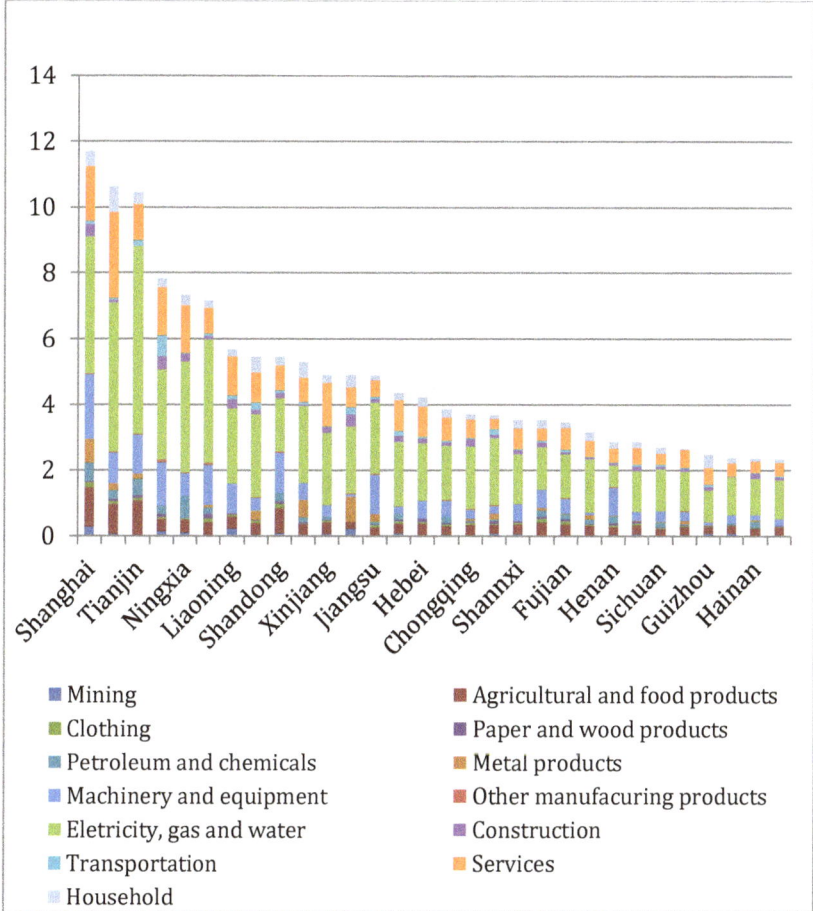

Legend:
- Mining
- Clothing
- Petroleum and chemicals
- Machinery and equipment
- Eletricity, gas and water
- Transportation
- Household
- Agricultural and food products
- Paper and wood products
- Metal products
- Other manufacuring products
- Construction
- Services

Conclusion and Future Work

A comprehensive multi-regional input-output analysis is able to capture embodied CO_2 emissions along the whole intra- and inter-regional supply chain. Especially, a country like China with a huge disparity between the wealthy coastal regions and relatively poor central and western regions needs such detailed accounting and modeling that can be used for climate change mitigation policy design and long-term development plans in terms of balancing economic growth and achieving emissions reduction targets. In this study, we mapped the embodied emissions in the inter-regional supply chain and demonstrated that the large domestic final consumption and international exports in the coast regions lead to huge amounts of CO2 emissions across China. Therefore, a continuing improvement of living standards in the poor regions of China may potentially cause a huge amount of emissions within the region and further impacts on the emissions in other regions.

Further readings

Guan, D., & Hubacek, K. (2010). China can offer domestic emission cap-and-trade in post 2012. *Environmental Science & Technology*, *44* (14), 5327-5327.

Guan, D., Hubacek, K., Weber, C., Peters, G., & & Reiner, D. (2008). The drivers of Chinese CO2 emissions from 1980-2030. *Global Environmental Change*, *18* (4), 626-634.

Guan, D., Peters, G., Weber, C., & Hubacek, K. (2009). Journey to world top emitter: An analysis of the driving forces of China's recent CO2 emissions surge. *Geophysical Research Letters*, *36* (LO4709), 1-5.

Hubacek, K., Feng, K., & Chen, B. (2011). Changing Lifestyles Towards a Low Carbon Economy: An IPAT Analysis for China. *Energies*, *5* (1), 22-31.

Minx, J., Baiocchi, G., Peters, G., Weber, C., Guan, D., & Hubacek, K. (2011). A "Carbonizing Dragon": China's Fast Growing CO2 Emissions Revisited. *Environmental Science & Technology*, *45*, 9144-9153.

Peters, G., Weber, C., Guan, D., & Hubacek, K. (2007). China's Growing CO2 Emissions: A Race between Increasing Consumption and Efficiency Gains. *Environmental Science & Technology*, *41* (17), 5939-5944.

Chapter 11: Waste Flows in Multi-regional Input-Output Models

Christian John Reynolds and John Boland

Introduction

Waste is an umbrella term for items that no longer serve their intended purpose or function. In the present economy of the developed world, acts of production and consumption create waste. Every type of waste has its own waste cycle and an optimum method of disposal, from simple recycling for aluminium cans or food scraps to more complex processing for bio-hazardous material.

Yet, with such large amounts of waste being generated, there are few tools that give us an understanding of the environmental, economic and social impacts of waste at a national level, or of how these impacts resolve at a regional or local level. This lack of understanding of local and regional waste issues prevents the effective management of waste to the detriment of the environment and public health.

This chapter provides an introduction to the analysis of waste using a multi-regional waste input-output (MRWIO) model. It concludes with a case study on how to apply an MRWIO to the current Australian economy.

Waste Input Output analysis

Waste input output (WIO) analysis is used by researchers and practitioners to model the flow of waste types throughout the economy. It has applications ranging from water and sewage waste to 'smart waste' from manufactured goods.

Since the development of WIO in Japan in the 1990s[1], WIO has become recognised as an extension of IO analysis, as a form of Life-Cycle Assessment (LCA) and as a variant of Material Flow Analysis (MFA).[2]

Most WIO studies have focused on Japan where the WIO methodology was first developed. WIO tables were constructed for Japan in 1995 and 2000[3]. Japan's 1995 WIO table originally comprised 78 industries, 24 waste types and 9 types of bulky waste, which were then condensed into 13 industries, 13 wastes and 3 treatment processes.[4]. This national model was then disaggregated into an MRWIO for nine regions of Japan.[5] The Japanese MRWIO has since been expanded to include 32 household and 18 industrial waste types.

Figure 1 and Table 1 represent two methods of displaying the waste flows of an economy. Figure 1 represents the flows of waste (solid lines), goods, services and capital (dotted lines) and recycled goods re-entering the economy (dashed lines). Figure 1 also splits landfill, energy recovery and recycling from other industrial sectors to illustrate where the waste is being sent for treatment or disposal. This is a holistic overview of waste flows with little opportunity to understand the 'black box' of each part of the waste system.

In tables below, the grey shaded areas represent the parts of the IO table that have been added to the standard table by the introduction of waste data.

Table 1 displays the interactions of waste and the economy in an input output (IO) format that allows the reader to track the individual waste flows. The WIO table separates the waste treatment sector (T_{12}) of the economy into the various treatment types available to that economy (e.g. recycling, landfill, incineration). The waste treatment sectors are placed as columns to the right of the rest of the economy. The WIO disaggregates the waste stream into various waste types that the modeller wishes to simulate. The amount of waste produced by each industry is placed below that industry's column in a horizontal 'satellite' row at the bottom of the IO table. The cells in this row are denoted as either W_{31}, waste produced by the intermediate sectors of the economy (T_{11}), or as W_{32}, waste produced by the waste industry (T_{12}), which partly represents the waste produced by the treatment of waste.

[1] Nakamura, S. (1999). Input-output analysis of waste cycles. Environmentally Conscious Design and Inverse Manufacturing, 1999. Proceedings. EcoDesign '99: First International Symposium

[2] Kagawa, S., S. Nakamura, et al. (2007). "Measuring spatial repercussion effects of regional waste management." Resources, Conservation and Recycling **51**(1): 141-174.

[3] Published online at: http://www.f.waseda.jp/nakashin/wio.html (accessed 01/11/2012)

[4] Nakamura, S. and Y. Kondo (2002). "Recycling, landfill consumption, and CO2 emission: analysis by waste input–output model." Journal of Material Cycles and Waste Management **4**(1): 2-11,

[5] Kagawa, S., H. Inamura, et al. (2004). "A Simple Multi-Regional Input Output Account for Waste Analysis." Economic Systems Research **16**(1): 1 - 20, Kagawa, S., S. Nakamura, et al. (2007). "Measuring spatial repercussion effects of regional waste management." Resources, Conservation and Recycling **51**(1): 141-174..

Figure 1: A Pictorial mapping of economic and waste flows in a hypothetical economy

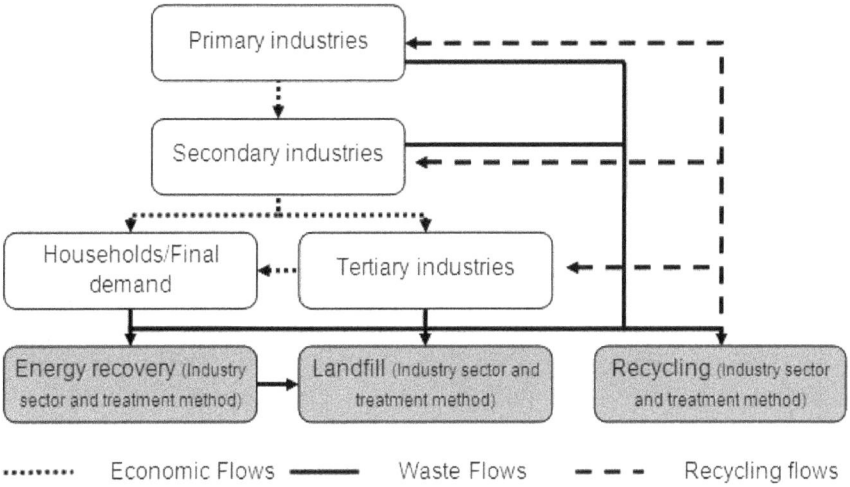

Table 1: The Waste Input-Output model.
Notation as follows:
- T_{11} Intermediate demand transaction matrix
- T_{12} Waste treatment sectors transaction matrix with other Intermediate sectors
- F Final Demand transaction matrix with transaction Intermediate sectors
- x_1 Gross output matrix of Intermediate supply sectors
- W_{31} Waste production from intermediate sectors
- W_{32} Waste production from Waste Treatment sectors
- W_F Waste production from Final demand
- x_3 Total waste production

		Intermediate demand	Waste treatment (Wt)	Final demand (F) (and exports)	Gross outputs
		Sector 1... N_1	Sector 1...N_2	Sector 1...N_f	
Intermediate supply	Sector 1 ... N_1	T_{11}	T_{12}	F	x_1
Waste generation by type (Wg)	Sector 1 ...N_3	W_{31}	W_{32}	W_F	x_3

Prior to the Japanese WIO approach, other methods for waste analysis in IO frameworks existed under the umbrella term "Environmental IO" analysis. These had the fundamental problem that they allowed each waste classification to have only one method of treatment. This is not the reality of modern waste flows, where different amounts of waste within one waste classification are recycled, incinerated for energy or sent to landfill and so on.

The WIO model allows for multiple treatment methods, providing a separate table (or matrix) that gives information on where different percentages of the waste are allocated for treatment. With the previous Environmental IO methodologies, food waste could be treated by sending to landfill or, composting but not both concurrently. Under the WIO framework different components of food waste can be sent to each treatment. To understand how the allocation matrix operates we can examine the numerical example in Tables 2 and 3.

Tables 2 and 3 are hypothetical WIO tables that only show the intermediate sectors, the two good producing sectors ($N_1=2$), four waste treatment sectors ($N_2=4$) and two waste types ($N_3=2$).In order to perform IO analysis, Table 2 needs to be square and symmetric (with the same number of corresponding columns and rows). However, because we have 2 waste types and 4 treatment processes Table 2 is not symmetric - rather it is a 6×4 table, therefore it has to be converted through the use of an 'allocation' matrix.

Table 2: Hypothetical numerical example of an asymmetrical WIO system displaying both waste type and treatment method.

	Good 1	Good 2	Treat. 1	Treat. 2	Treat. 3	Treat. 4
Good 1	0.09	0.15	0.05	0.01	0.03	0.16
Good 2	0.09	0.03	0.09	0.06	0.14	0.48
Waste type 1	0.06	0.14	0	0	0	0
Waste type 2	0.16	0.05	0.4	0.3	0	0

Table 3: Hypothetical numerical example of a symmetric WIO system, displaying waste flows only by treatment method.

	Good 1	Good 2	Treat. 1	Treat. 2	Treat. 3	Treat. 4
Good 1	0.09	0.15	0.05	0.01	0.03	0.16
Good 2	0.09	0.03	0.09	0.06	0.14	0.48
Treatment 1	0.033	0.077	0	0	0	0
Treatment 2	0.027	0.063	0	0	0	0
Treatment 3	0.144	0.045	0.36	0.27	0	0
Treatment 4	0.046	0.075	0.04	0.03	0	0

The difference between Tables 2 and 3, is that the rows of waste types (Table 2) have been replaced by rows of Treatments (Table 3). It is important to understand that this means that table 2 indicates that Good 2 produces 0.14 of Waste Type 1 as part of its production process. While Table 3 informs us that Good 2 has 0.077

of its waste treated by treatment method 1. The different information on the two tables allows the reader a greater understanding of the waste flows.

Table 4 is termed in WIO literature as the 'allocation' matrix (labelled mathematically as $\boldsymbol{S}_{23}$). This matrix informs the reader where each waste type is to be treated. Table 4 indicates that 55% of Waste Type 1 flows to Treatment 1, 45% to Treatment 2 and 5% to Treatment 4. For Waste Type 2, 90% is allocated to Treatment 3 and 10% to Treatment 4. Modifying these percentages changes the amount of waste flowing to that treatment type.

Table 4: The associated allocation matrix for Tables 2 and 3. This table indicates what amount of each waste type is treated by which treatment method. In mathematical notation this matrix is referred to as S_{23}.

	Waste type 1	Waste type 2
Treatment 1	*0.55*	0
Treatment 2	*0.45*	0
Treatment 3	*0*	0.9
Treatment 4	*0.5*	0.1

Table 4 is used to convert Table 2 into the symmetric Table 3. This is achieved by multiplication from the left by the allocation matrix, sorting the generation per waste type (food waste, plastics, metals etc) into the amount of waste treated per type of waste treatment services (compost, landfill, recycling etc) as shown in Figure 2. In the now symmetric 6×6 Table 3, the waste components are a 4×6 sub-matrix, with the columns sorted via treatment method not waste type.

Figure 2: The mathematical conversion of Table 2 into Table 3, via multiplication from the left by the allocation matrix $\mathbf{S}_{2\,3}$

$$\overbrace{\begin{bmatrix} 0.55 & 0 \\ 0.45 & 0 \\ 0 & 0.9 \\ 0.5 & 0.1 \end{bmatrix} \times \begin{bmatrix} 0.06 & 0.14 \\ 0.16 & 0.05 \end{bmatrix}}^{\mathbf{S}_{23}\mathbf{W}_{31}} \quad \overbrace{\begin{bmatrix} 0.55 & 0 \\ 0.45 & 0 \\ 0 & 0.9 \\ 0.5 & 0.1 \end{bmatrix} \times \begin{bmatrix} 0 & 0 & 0 & 0 \\ 0.4 & 0.3 & 0 & 0 \end{bmatrix}}^{\mathbf{S}_{23}\mathbf{W}_{32}}$$

$$= \begin{bmatrix} 0.033 & 0.077 & 0 & 0 & 0 & 0 \\ 0.027 & 0.063 & 0 & 0 & 0 & 0 \\ 0.144 & 0.045 & 0.36 & 0.27 & 0 & 0 \\ 0.046 & 0.075 & 0.04 & 0.03 & 0 & 0 \end{bmatrix}$$

WIO analysis applied to Australia

An MRWIO for Australia requires substantial collection of information on waste disposal and creation. This information ranges from the categories and volumes of waste produced by industry sectors and households, to the waste treatment options available in each region. In Australia, the Australian Bureau of Statistics (ABS) and state based environmental agency counterparts regularly publish data on the waste production of the Household (Municipal (MSW)); Commercial and Industrial (C&I); and Construction and Demolition (C&D) sectors[6]. Unlike in Japan, the large geographic distances between Australian regions (states and territories) have resulted in less than 5% of waste being moved between regions until after treatment. This lack of cross-state waste flow allows the use of national reports (such as the *National Waste Report 2010*) in addition to state-based data sets, for MRWIO construction.

Table 5: Non Symmetric single region layout with 3 waste types (N_3=3) and 3 treatment sectors (N_2=3)

	Sector 1 ... Sector n	Compost	Recycle	Landfill	Final Demand	Total
Sector 1 ... Sector n	T_{11}	T_{12}			F	x_1
MSW	W_{31}	W_{32}			W_F	x_3
C&I						
C&D						

The Australian waste and input-output data are arranged in the format of an IO table, with the volumes of waste produced by industry sectors or treatment methods displayed in Table 5 as W_{31} and W_{32} respectively. Tables 5 and 6 present a schematic for a single state or region of Australia in non-symmetric and symmetric formats, with the waste types and treatments shaded in grey.[7]

---

[6] Australian Bureau of Statistics (2010). 4613.0 Australia's Environment Issues and Trends Special issue: Climate Change; Australian Bureau of Statistics (2010). 4602.0.55.002 Environmental Issues: Waste Management And Transport Use; (2010) National Waste Report 2010, Environment Protection and Heritage Council.

[7] Though hypothetical in this chapter, a full MRWIO of Australia could be based on the Australian Bureau of Statistics Input output tables (5209.0.55.001 - Australian National Accounts: Input-Output Tables, 2008-09). These have 112 industry sectors (N_1=112) in

Table 6: Symmetric single region layout with 3 waste types (N_3=3) and 3 treatment sectors (N_2=3)

	Sector 1 ... Sector n	Compost	Recycle	Landfill	Final Demand	Total
Sector 1 ... Sector n	T_{11}	T_{12}			F	x_1
Compost Recycle Landfill	$W_{23}W_{31}$	$S_{23}W_{32}$			$S_{23}W_F$	$S_{23}x_3$

Table 7: Schematic for an Australian MRWIO Table with 8 regions. Highlighted and shaded areas represent multiple placements of Table 6, with the rest of the table cross regional flows.

addition to the three waste creation (N_2) (Domestic and Municipal, Commercial and industrial, and, Construction and Demolition), treatment (N_3) (composting, recycling or landfill) sectors (N_2,N_3,=3), and N_F final demand sectors.

The full Australian MRWIO schematic (Table 7) has eight highlighted and shaded areas that represent multiple placements of Table 6. Each of these variants of Table 6, represents a state of Australia (New South Wales (NSW), Victoria (Vic), Queensland (QLD), Western Australia (WA), South Australia (SA), Tasmania (Tas), Northern Territory (NT), or the Australian Capital Territory (ACT)), while the dashed cells of Table 7 represent cross-regional flows of dollars or waste.

Application of WIO

A WIO table displays the waste and monetary flows through an economy. It can be used to:

- Explore the value and volumes of waste going to landfill.
- Examine the effects of recycling on the economy, along with the identification of the mass, type and value of recycled materials.
- Understand which regions have active recycling industries and plan waste strategies accordingly.

Furthermore, once a WIO table has been produced, *direct* and *indirect waste generation coefficients* and the *Leontief multiplier* can be calculated[8]. These can inform both industry as well as government planning in a number of ways:

- The tonnages of waste directly generated for each dollar invested in an industry sector can be calculated, along with the waste tonnages 'indirectly' generated upstream due to this economic activity. These *direct* and *indirect waste generation coefficients,* and the *Leontief multiplier* allow the industry sectors that generate waste with the highest intensity (the most waste per dollar spent) to be identified and (policy) action taken to either reduce the waste produced or treat (recycle) the waste more efficiently.[9]
- Hypothetical scenarios can be run where changes to either the volume of waste produced (alterations to the MRWIO table), or to the treatment method used (alterations to the allocation matrix) allow practitioners to observe the effects of differing waste management strategies upon waste flow and the economy.

---

[8] The creation of these multipliers is highly technical. For this reason further detail is not included in this chapter. Refer to Nakamura, S. and Y. Kondo (2009) for a worked introduction to their creation and utilization.

[9] Direct tonnages are the tonnages created by activity solely within the sector in question, found via the coefficient multiplier. Embodied tonnages are the total tonnages that are created throughout the entire economy, via activity within the one sector, found via the Leontief multiplier.

Conclusion

This chapter has provided an overview of the modelling approaches and the issues surrounding the implementation of a Waste Input-Output (WIO) and Multi Regional Input-Output (MRIO) model in Australia. This chapter has outlined some of the challenges in creating a contemporary integrated MRWIO for Australia, as well as giving a review of the history and theoretical construction techniques of WIO

The data requirements of the combined MRIO and WIO model are complex, drawing not only upon traditional IO tables but also on sources that relate to the differing treatment methods used in each region and the levels of waste produced in every sector of the economy. Because of limited availability, some regional WIO table entries have to be estimated from national data. Yet this should not deter potential creators or users of MRWIO accounts. The rewards for such a comprehensive mapping of waste within an economy are great. MRWIO tables provide us with the ability to identify waste producing industries through the supply chain and model the impacts of changes to waste policy.

Further reading

Kagawa, S. (2012). Frontiers of environmental input-output analysis. New York, NY: Routledge.

Kagawa, S., Inamura, H., & and Moriguchi, Y. (2004). A Simple Mult-Regional Input-Output Account for Waste Analysis. Economic Systems Research , 16 (1), 1-20.

Nakamura, S., & Kondo, Y. (2009). Waste Input-Output Analysis: Concepts and Application to Industrial Ecology. New York: Springer.

Acknowledgements

Many thanks to the team at ISA in the University of Sydney for all their support and comments on this chapter. The research connected with this chapter was supported in part by the Waste Management Association of Australia Young Professionals scholarship, and the ARC Linkage Project 'Zeroing in on Food Waste: Measuring, understanding and reducing food waste' (LP0990554) funded by the Australian Research Council and industry partners Zero Waste South Australia and the Local Government Association of South Australia. Early drafts of this chapter were written as part of the assessment towards certification at the International School of Input-Output Analysis.

Chapter 12: An Enterprise MRIO for a University

Christopher Dey, Sandra Harrison, Manfred Lenzen, and Joy Murray

Introduction and Background

This chapter describes the development, implementation and initial outcomes of an enterprise level multi-regional input-output (MRIO) model and planning tool for the University of Sydney. The MRIO tool, known as the University Sustainability Model (USM) for ease of reference, is the result of an internal collaboration between researchers in the Integrated Sustainability Analysis (ISA) research group, the Planning & Information Office (PIO) as well as Information Communication & Technology (ICT).

Most universities are very large and complex organisations. Given their various functions (teaching, research, and community engagement), as well as their many stakeholders (individuals, governments and business) they are arguably much more complex to manage than other organisations of similar overall size in terms of employees or turnover. About five years ago senior financial and strategic planning staff in the University saw the potential of ISA's research for gaining an understanding of the underlying and sometimes opaque consequences of the University's operations. They wanted to improve performance reporting across the University's many portfolios and functions and, perhaps most importantly, to add clarity to informed decision making on resource allocation.

Previous applications of input-output techniques to organisations had mainly examined the environmental performance of institutions in terms of their draw on external resources, for example, in terms of the ecological footprint of an institution or its components. Previous work had also extended such single-entity analysis to cover many indicators in a triple bottom line approach. Whilst these approaches were useful in understanding overall sustainability impacts of the University and for reporting, they were too aggregated to assist with more detailed management decisions. Hence given the many interdependencies within universities, there was a clear need for a comprehensive enterprise input-output model that could also accommodate the *internal* economy of the university.

The Enterprise Model

MRIO Framework

Although enterprise models themselves are not new, the USM is novel in 1) its large scale; 2) its incorporation of both Leontief and Ghosh analytics; 3) the consideration of system closure and 4) analytical procedures such as structural path analysis (SPA).

In the USM MRIO, region 1 represents the University's interdependencies between 'organisational units' (sectors), and region 2 is the Australian economy (Figure 1). With respect to the rest of the world the model is open: exports are part of final demand and imports are a primary input. A key aspect of balancing such a system, particularly the $\mathbf{T}_{UU}$ block, is dealing with increases and decreases in inventories. Rather than taking a commodity view of inventories, we regard 'increases in financial reserves' as a primary input and 'decreases in financial reserves' as a component of final demand, with both always of positive value.

	University	Australia	Final demand
University	$\mathbf{T}_{UU}$	$\mathbf{T}_{UA}$	$\mathbf{y}_U$
Australia	$\mathbf{T}_{AU}$	$\mathbf{T}_{AA}$	$\mathbf{y}_A$
Prim.input	$\mathbf{v}_U$	$\mathbf{v}_A$	o

Figure 1: Structure of the USM, with subscripts U and A referring to the University and Australian economies respectively. There are various components of final demand (y) and primary inputs (v). The dotted line separates components of final demand, for example private final consumption, or expenditure on gross fixed capital. Figure from Lenzen et al (2010).

The main population of the USM results from an extraction of University account data (general ledger). Revenue transactions were used to construct $\mathbf{T}_{UU}$, $\mathbf{T}_{UA}$ and $\mathbf{y}_U$. Expenditure transactions were used to construct $\mathbf{T}_{AU}$ and $\mathbf{v}_U$. The procedure, described here in brief terms, involved making a concordance between the

University's financial system class codes, detailed descriptions of all expenditures and revenues, and the Australian input-output classification (here industry by industry) that ISA has created (344 sectors, for example in T_{AA}). Careful matching is required of pairs of internal expenditures and revenues, and T_{UU} was balanced using a sophisticated optimiser.

Various sizes of the University economy were created, with the most detailed model comprising 16 faculties and a further 60 administration and support units. Each of these areas was further broken down into 17 activities, meaning the most detailed USM had 1292 sectors (76 x 17). Including the Australian economy in supply-use form, the full USM comprises approximately 2000 sectors. Hence for testing and demonstration purposes a smaller USM version was constructed, containing the 16 faculties + 2 administration units, each separated into teaching, research and 'other' activities, making a total of 54 sectors within the University. One of the key financial flows in the University concerns the allocation to different organisational units of untied Federal Government funding. These flows are an important element of University management and politically sensitive. Incorporation of the actual University accounts from a previous year, rather than projections or estimates is important for the credibility and acceptance of such a tool.

Satellite Accounts

Whilst the reclassification of the University general ledger from a single accounting system largely defines the USM, constructing the various satellite accounts for the USM involves many other data sources and their separate systems. Generally, these data sources are in a very different classification structure. For example, annual energy, water use, waste and emissions data on campus typically cover several buildings, each of which may be occupied by several faculties (and perhaps other organisational units). Hence a disaggregation and reclassification exercise is needed to populate the satellite data to align with the USM classification. Detailed building occupancy data, including room uses (for example teaching, research and administrative space) assist with this task.

In addition to the environmental satellite data, other performance data for the University were incorporated in the satellite accounts. These included data on employment (split into academic and general staff), student load data (undergraduate and post graduate), degree completions, various publication data and other research metrics. Most of these data could be easily mapped to the USM structure, but each data source required separate concordances and data handling procedures.

Implementation

Requirements

From the outset, the intention was that the USM, as incorporated into a tool, had to be able to be operated by non-specialist input-output practitioners. In planning the tool development, careful consideration was given to the USM being:

- developed so that it met the business requirements of the University
- easily operated and explained by non-experts
- practical and relevant – useful for making important decisions
- robust, in the sense that the software tool must be in a 'production' environment, rather than an unstable prototype available on a limited number of computers
- updated and maintained by non-experts
- integrated with other planning, reporting and management tools within the University.

Approach

The ISA research group had previously collaborated with software developers to create a software tool for performing triple bottom line and life cycle analyses of organisations. This web-based tool was expanded and adapted to accommodate the much larger and more complex USM system. Key considerations were ease of use, reliability, ability to accommodate ongoing USM data updates and expansion, and computational time. This last point was not trivial. In principle, several users could be calculating the inverse of a full USM model of ~2000 sectors at the same time. Effort was made to both improve basic calculation routines, and to avoid conflicts from multiple users' calculation times.

Practical implementation also required careful coordination of project team members with quite different backgrounds and expertise. ISA researchers were the IO specialists but had to collaborate with data source experts across the University with little IO background. The main 'client', at least initially, was the Planning and Information Office, who were expected to be the main user of the USM tool for conducting analyses and reporting. Software developers external to the University were engaged as experts on the actual tool and its functionality during the project development phase. Finally, the University's ICT portfolio was required to host and support the hardware and software environments for the development and ongoing use of the USM tool. Active and sensitive project management of all of these parties and issues, to a tight budget, was a key element of the project. Underpinning this project approach was the principle of ensuring all development reflected the business requirements of the University.

Screenshots

The USM tool is still under final development, so we include here some representative screen shots only.

Input Screens

On the input side, the USM tool requires the user to make several choices for each *analysis[1]*. For the analysis of interest, different combinations of components of final demand can be chosen (Figure 2). There are myriad possible combinations of these, but most are not sensible or of interest. For example, an analysis may be defined by Private final consumption (households) expenditure of teaching from a particular faculty. This source of revenue to the University represents domestic student fees. Another analysis might look at total final demand for all teaching and all research for a particular faculty. The next choice to be made concerns the definition of a *benchmark* or comparison analysis. The procedure is similar to the analysis choice, but usually involves a larger aggregation of the organisational units. For example, a teaching benchmark might comprise all final demand components of all organisational units (mainly faculties). An institution-wide benchmark is simply created using total final demand for the University. On a practical level, neat software functionality that allows quick expansion and collapsing of the USM sector hierarchy is very useful.

In terms of *indicators*, the USM allows the user to select various elements of the satellite accounts for which results (such as total impacts, total multipliers and so on) can be calculated. Some of the indicators are University-specific (such as degree completions) but others are common, and thus include contributions from both the internal and external economies (such as energy use or direct emissions).

---

[1] Even the choice of wording for the basic function of defining the 'entity' for calculations (organisational unit, or activities within them) was not agreed unanimously amongst all parties in this collaboration.

Figure 2: Screenshot showing in this case an analysis of all the activities within the organisational unit of the Faculty of Dentistry. Different components of final demand can be chosen (ticked). The activity descriptions are derived from the account codes.

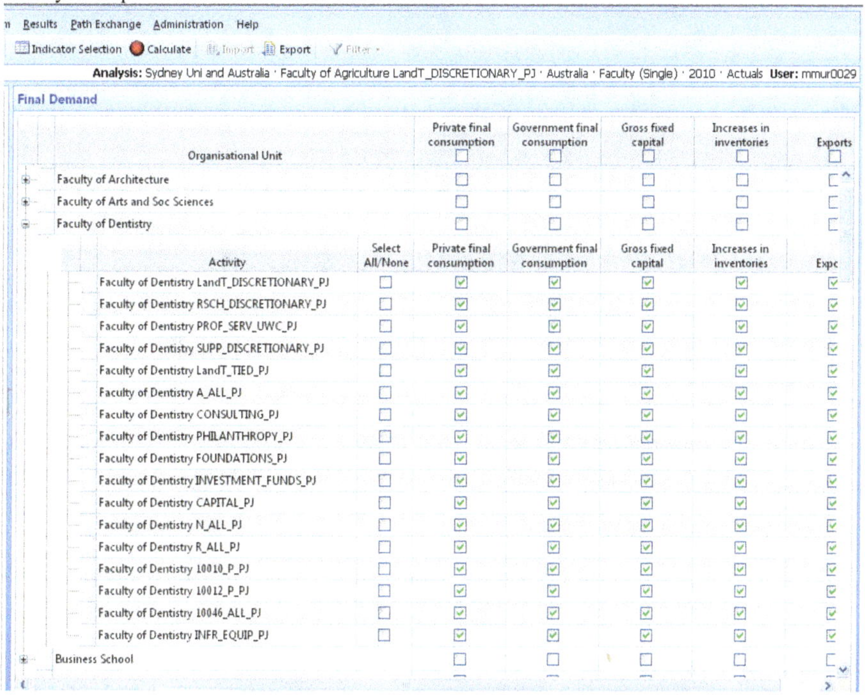

Result Screens

For illustrative purposes only, a sample results screen is shown in Figure 3. These results are the total impacts (or total requirements, or footprint), occurring in both the University economy and the Australian economy, that result from the total final demand of teaching in a single faculty, a value of 1,693 hectares. For comparison the benchmark analysis, which in this case is the ecological footprint of the entire University, is over 50,000 ha.

Figure 3:   Total impacts (highest level results) for the analysis case of the Faculty of Agriculture and Environment's teaching activities. The first indicator here is ecological footprint. Total impacts for nine sub-indicators are also shown.

The relative performance of the analysis case against the benchmark sector can be efficiently shown using a so-called spider or radar plot (Figure 4).

Figure 4:   Radar plot showing the normalised ecological footprint and its components of an analysis case (Faculty of Agriculture and Environment teaching) against that of the whole University. Results are normalised by dividing by the final demand total. Data points on the inner polygon labelled 0.1 would mean that the analysis case is 10 times lower in ecological footprint per $ of final demand than the overall University. Uncertainty levels are also indicated.

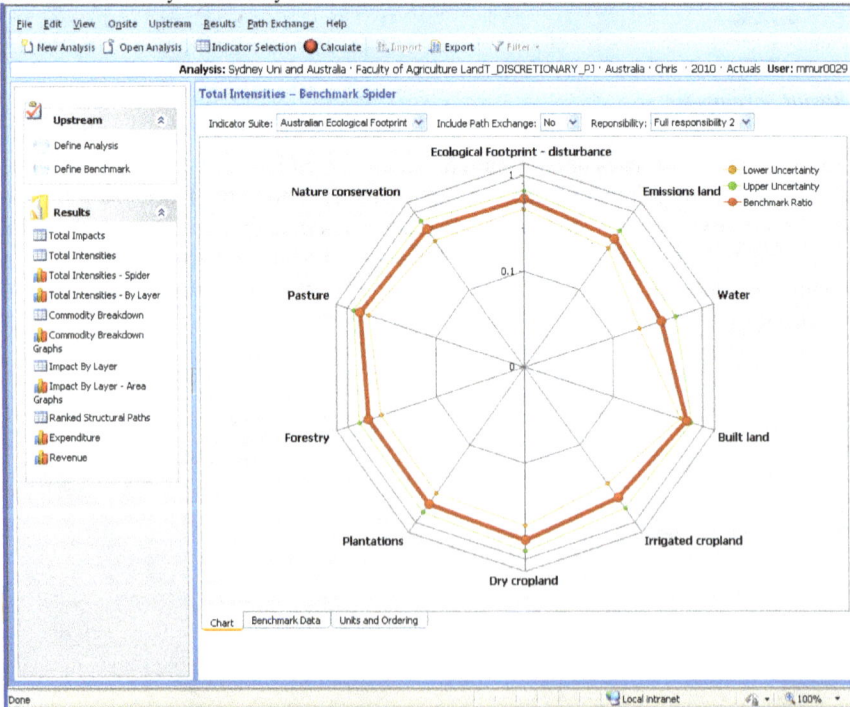

Discussion

At the time of writing, the status of the USM is that:

- the principles of the USM have been firmly established
- the core software has been developed and deployed in a production environment
- a procedure for near automatic updating of the main transactions matrix in the USM has been demonstrated, based on the University accounts
- connections made and preliminary data compiled for many satellite data to be incorporated
- USM user guide (manual) and update guide have been established
- there is preliminary buy-in from University portfolios which will ensure successful implementation at a business level.

To date however, there has not been extensive analysis and reporting of results based on the use of the USM. This has largely been a resourcing issue, always a risk in project implementation, and not a technical restriction of the USM. Our experience does suggest that engagement with potential end users and a tangible link to business requirements is essential for successful development. In addition, further training and upskilling of a broad range of staff is required in order that more users become sufficiently familiar with the technical operation of the USM tool and interpretation of results. Simple distribution of USM results, without appropriate context and insight, could lead to confusion at best and cynicism/scepticism at worst

The approach taken here is an example of true unified reporting for an enterprise: a single system can be used to report performance across a range of indicators, for many sub-entities. The calculations for each indicator are done in the same way (the single system) and cover all contributions, both direct and indirect. Already the USM has been used to confirm what is effectively cross-subsidisation of research activities by teaching income – a situation that is likely to be common to all research-intensive universities in the Australian higher education sector. Whilst this notion was not a surprise, the USM allows quantification of these effects, done in a systematic and repeatable way.

In terms of further developments, annual updating of the model data is already taking into account changes in the chart of accounts, such as new University entities and new classification codes. These changes are generally incremental and can be accommodated. On a broader level within the University, much effort is being spent on integrating the University's many data sources and their systems into a more unified data framework. In the past, quite separate IT systems have existed for major administrative functions such as financial accounting, human resources management, international relations, philanthropy, student data management, research management, utilities reporting and infrastructure management. Full integration of all such systems is a major undertaking, but in the near term at least a common reporting framework can be established to complement existing strategic and granular reporting tools. This will greatly assist the compilation of many of the satellite data for the USM.

Indeed, by its very nature the USM represents a highly capable framework for connecting these data sources.

Conclusions

In operationalising the USM MRIO framework into a broadly useable tool to enhance reporting capabilities and to support improved strategic decision making within a large organisation, we have learnt the following:

- Software development takes time: it is easy to underestimate the resources required to create complex but powerful, user-friendly tools
- Some IO concepts must be conveyed to users: users must appreciate some basic IO concepts
- Language is important: creating an IO-based tool for a general audience requires careful attention to language and jargon
- A partnership approach is required: all parties (IO experts, code developers, IT infrastructure people, data custodians, analysts and tool users) must appreciate each other's roles and work together
- As with all project implementation, business engagement and a clearly identified project 'champion' are essential.

Further reading

Baboulet, O. and M. Lenzen (2010). Evaluating the environmental performance of a university. *Journal of Cleaner Production* **18**: 1134-1141.

Foran, B., M. Lenzen, C. Dey and M. Bilek (2005). Integrating sustainable chain management with triple bottom line accounting. *Ecological Economics* **52** (2): 143-157.

Lenzen, M., C. Benrimoj, and B. Kotic (2010). Input-output analysis for business planning: a case study of the University of Sydney. *Economic Systems Research* **22** (2): 155-179.

Lenzen, M., B. Gallego and R. Wood (2009). Matrix balancing under conflicting information. *Economic Systems Research* **21** (1): 23-44.

Wood, R. and M. Lenzen (2003). An application of an improved ecological footprint method and structural path analysis in a comparative institutional study. *Local Environment* **8** (4): 365-386.

Acknowledgements

The authors would like to acknowledge gratefully the contributions to this tool and project made by University of Sydney colleagues Joseph Benjamin, Alison Byrne, Darren Dadley, Ammar Elnour, Christina Fava, Keith Jebb, Kim Koorey, and Joel Turner, as well as other staff in the Planning & Information Office, Information Technology & Communication and Campus Infrastructure Services. We would also like to thank the Capiotech software developers Stephen Pennisi and Matthew Vale for excellent technical implementation.

Chapter 13: Water Footprints for Spanish Regions Based on a Multi-Regional Input-Output (MRIO) Model

Ignacio Cazcarro, Rosa Duarte, and Julio Sánchez-Chóliz

Introduction

The increasing social awareness of our responsibility to understand and reduce our impact on the planet has prompted the search for tools to quantify the human demand on natural resources. Inspired by the ecological or footprint concepts, Virtual Water (VW) and Water Footprint (WF) have been used to measure the extent of human appropriation of this natural asset in our consumption of goods and services. The insights obtained from this concept result in a comprehensive analysis that moves from the traditional producer responsibility view towards a consumer responsibility approach that also considers the role of final consumption as an important driving force for water demand. The WF can be measured by calculating the water resources required for the production of the goods and services in an area, minus the part of those resources that were embodied in exports (i.e. VW exports), plus the embodied water contents of imports (i.e. VW imports). Using these concepts, the dependence on water imports or the net export of VW of countries (or regions or river basins) can be studied. It is also possible to evaluate the coherence and efficiency of water uses, management and allocation.

In this context, water footprints have also gained importance in the policy field. For example, Spain has included the WF in its water planning. Ministerial Order ARM/2656/2008 included the Water Planning Instruction for the development of new management plans in Spain, adapting the Spanish law to the EU Water Framework Directive (2000/60/EC).

The Ministerial Order states that Basin Water plans must include information of the "different economic activities that affect water use, providing aggregate data for the river basin and, where appropriate, for the regional level". In this regard, "there will be an analysis of the water footprint of different socio-economic sectors, defined as the total domestic water and net water balance of imported and exported water". As we will see, MRIO models allow practitioners to calculate the water impact associated with the current regional economic structure, WFs, as well as possible impacts resulting from different regional economic policies or development strategies. Since Water Planning in Spain is national, but driven in coordination with the regional Water Agencies, the disaggregated and multiregional character of MRIO models make them particularly suited to the analysis of regional environmental and socio-economic effects of the implementation of new Water Plans, Irrigation Plans and water policies in Spain.

For example, we may observe how demand in the largest metropolitan area, Madrid, has significant impact on scarce water resources in the regions of Andalucia or Murcia. The Aragonese region suffers from water pollution, which means a reduction of water availability for all users. This is an outcome of individual decisions regarding the location of pig farms, the product of which generates revenue that flows out of Aragon to neighbouring regions. These and other environmental pressures should be analyzed in terms of final demands, trade and all the stages of production to help design sustainable environmental policies. If we were to simply focus on the first steps of production and direct water consumption in an area that is mainly concerned with agriculture, we would overlook how pressure on water has increased. These pressures include the growth of population and tourism, which is also significant in the aforementioned regions; the development of water-intensive industries such as the textile and paper production; and changes in production orientation.

The Spanish case – motivation, input data and trade

Water security is a problem in Spain. There is considerable climatic variation between the North and the South and this is reflected in both temperature and precipitation. Rainfall can be erratic which results in recurring droughts in Central and Southern Spain. These facts, together with the recent introduction of water demand management perspectives in the Spanish Water Institutions and the emerging recognition of the WF as an informative tool in the design of water policy, have led to a growing interest in estimating the water embodied in domestic and traded goods among regions and abroad.

This chapter provides an example of the possibilities of MRIO models (see definition in chapter 1) to asses these water issues or other environmental pressures.

Input data and process

Following previous work on interregional matrices and IO models e.g. INTERTIO, we compiled trade and input-output data for the Spanish regions trying to get as much detail as possible in sectors especially relevant to the study of water. To be precise, we obtained data of 15 regions (Autonomous Communities), and estimated the three other regions lacking a regional table (grouped together here), distinguishing 40 sectors in each table. Trade data were deconstructed by the structural database C-intereg[1], which identifies the trade between regions (interregional, with a double entry table) for various types of goods. It gathers together the Permanent Survey of Road Goods Transport; the railway statistics on charging Unit and Combined Transport; the origin and destination matrices estimated from the maritime state port and air cargo movements information by the Ministry of Public Works; and the indirect estimated flows of energy products by pipeline obtained from the same Ministry.

Blue water is defined as the fresh surface and groundwater, in other words, the water in freshwater lakes, rivers and aquifers. Green water is defined as the rainwater stored in the ground and absorbed by the crops[2]. Despite not being mutually exclusive sets, both blue water and green water provide information about different water management and irrigation alternatives.

In this chapter, the two types of water data are combined with economic information to evaluate impacts. The pressure of Spanish households (citizens by region) on the global water resources comes from the domestic water consumptive use (WC), plus the VW in imports (VW^{imp}), minus VW in exports

$$(VW^{exp}): WF = WC + VW^{imp} - VW^{exp}$$

Interregional and international trade

Looking at heat map 1 we can see the whole table minimized so that the most important exchanges are coloured in black (in red in the online version), losing colour with the loss of relevance of the exchanges until white. Regional sales among sectors are represented in the main (intensively coloured) diagonal and in rows the exports to other regions and to the EU and RW. The columns represent the purchases by sectors (in the first line of each region, the agrarian sector, then 20 industrial sectors and 19 services) and imports from the EU and RW.

---

[1] About the C-Intereg database and methodology:
http://www.c-intereg.es/metodologia.asp#Ancla2.
[2] For the complete definition, see the WF network
(http://www.waterfootprint.org/?page=files/Glossary)

Heat Map 1: Spanish interregional, EU and RW trade of intermediates.

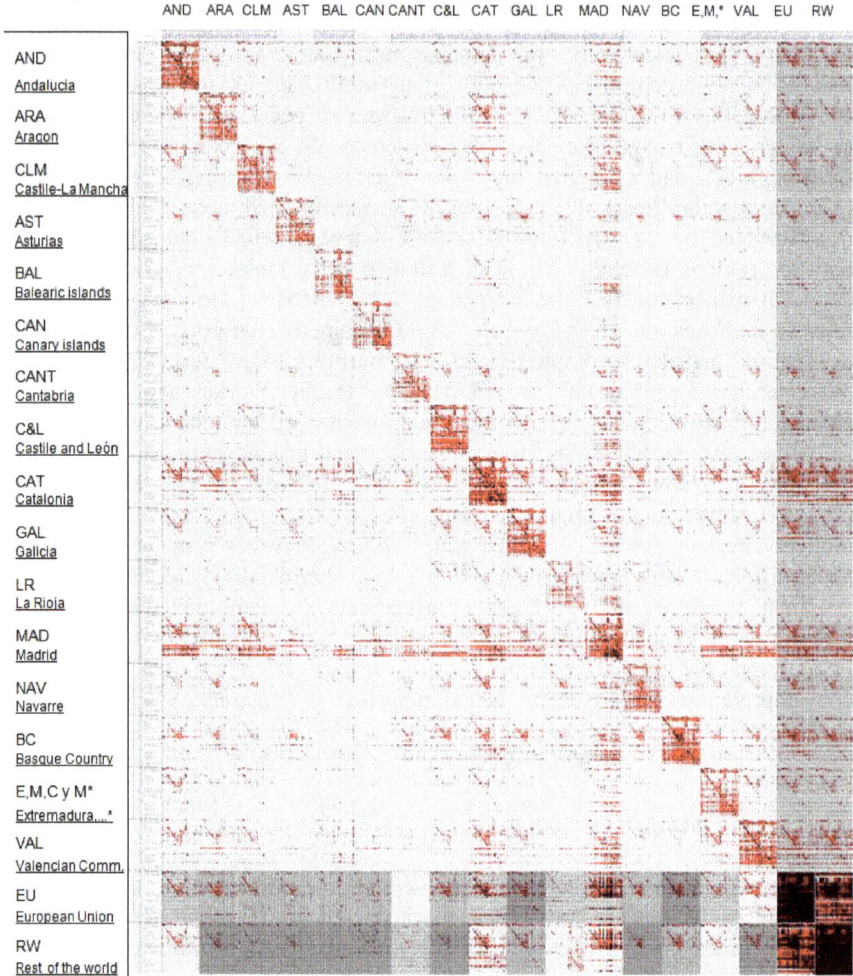

* Extremadura, Murcia, Ceuta and Melilla.
Scale of red color, showing more intense the high values (monetary transactions), and in almost black the highest ones.

Source: Own elaboration.

The interregional trade of Spanish regions is characterized by the importance of the regions of Catalonia and Madrid (which are the main economic drivers of the Spanish economy) and also Galicia and the Valencian Community. Madrid (and to a much lesser extent Catalonia) is the region where the lower rows of the 40 sectors represented for each region, which refer to services activities, are coloured in intense red. Looking at the columns, the EU and RW are also the main destination of goods and services followed by Madrid and Catalonia. As exporters, besides the obvious importance of the aforementioned regions (highlighting Madrid in the export of services), important regions are also the Basque Country (especially via industrial sectors), the Valencian Community (agrarian, textile, non-metallic minerals, etc) or a few industrial sectors (metallurgy, metallic and mechanical) of Asturias, a region with a significant mining and metallurgy tradition.

Results of the Water Footprint (WF) of Spain and within regions
Importance of the agrarian sectors and water consumption at point of origin

The size of the agrarian sector, in terms of value of production, is shown with the proportional size of the spheres in Map 1. Andalusia stands out at above 12,000 million €; above 5,000 million € are the regions of Castile and León and Catalonia; and around 4,000 million € are the regions of Murcia, Galicia, the Valencian Community and Aragon.

Map 1: Ratio of green to blue water and agrarian sector size (millions of €).

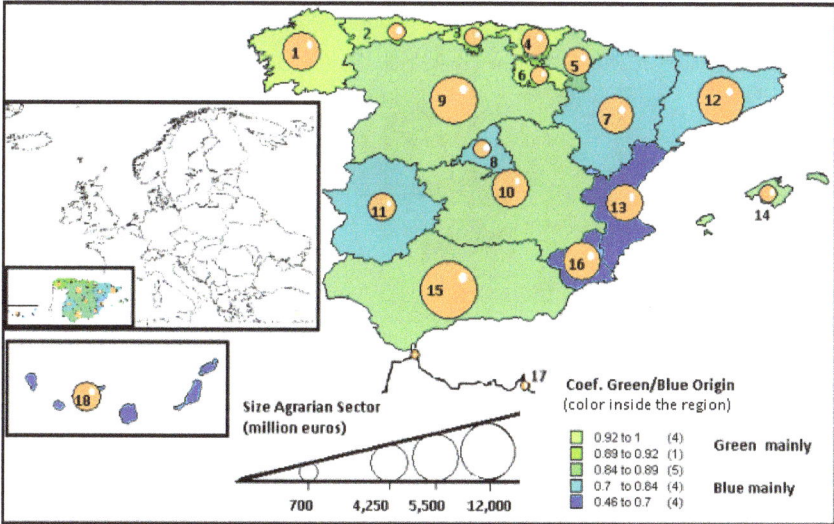

1. Galicia, 2. Asturias, 3. Cantabria, 4. Basque Country, 5. Navarre, 6. La Rioja, 7. Aragon, 8. Madrid, 9. Castile and León, 10. Castile-La Mancha, 11. Extremadura*, 12. Catalonia, 13. Valencian Community, 14. Balearic islands, 15. Andalusia, 16. Murcia*, 17. Ceuta and Melilla*, 18. Canary islands (Canaries). * Grouped together with one table for those regions.

The ratio of green to blue water is the highest in the north- west coast (lighter colour), then in the centre of the peninsula and the Balearic Islands. The lowest is in Extremadura, Madrid, Aragon, Catalonia and especially in the Valencian

Community, Murcia and the Canary Islands. Regulated water is in high demand in the eastern part of Spain where water availability is lower. Production in these regions is highly dependent on regulated resources, which have to be shared by several users (urban, industrial, agrarian, recreational activities, etc), and which often results in political rivalry for the resources.

VW trade and WFs by region

In Map 2 we observe the embodied (or virtual) water contents per € of final demand (which includes domestic and foreign demand) given by the size of the spheres. Agrarian goods, food industry, wood and cork, textile, restaurants, paper and printing, hotels, furniture, etc have the highest green water intensities This is common in many countries, but for many Spanish regions we also find high green water intensities in the sector of the arts, entertainment and recreation activity since, on one hand, it involves direct water consumption to maintain golf courses, football pitches and parks. On the other hand, the sector has high embodied water from the purchases of food and live animals for festivities, textile production, catering services, etc. The lighter the colour of the spheres in Map 2 (greener in the online version), the higher they are in terms of relative importance to the agrarian sector (with respect to other sectors) in the VW demand. The darker (blue) the spheres, the higher the relative importance of industrial and service sectors. The net regional exporters of VW are shown in white and light grey and net importers by darker grey until black. The main VW flow across regions is depicted by the arrows (in yellow in the online version) of Map 2.

Map 2: VW (Blue and Green) net exports and contents in Final Demand

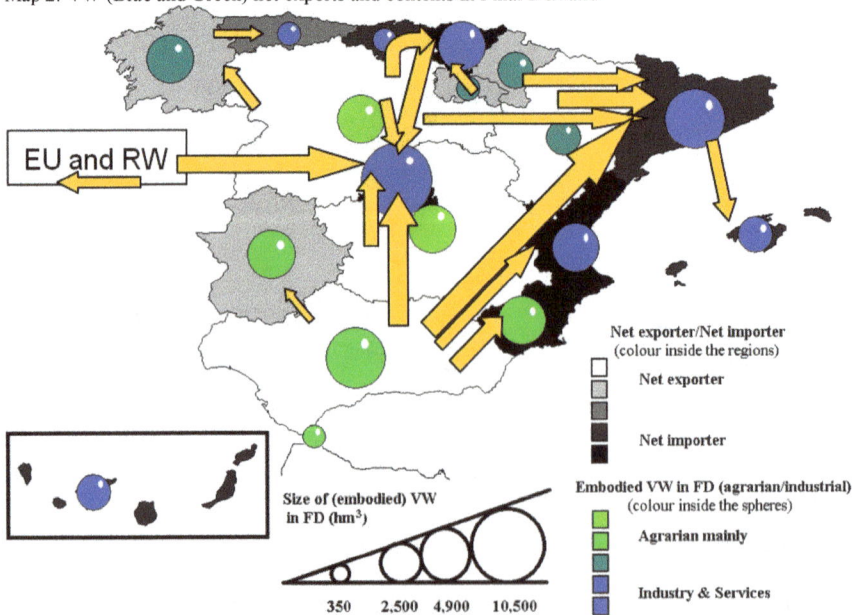

* Regions are colored according to their net export (positive values) or net import (negative) character
* The size of the spheres shows proportionally the size of VW contents in Final Demand.

* Greener implies a higher coefficient of VW from agrarian goods with respect to industrial.
* The arrows represented show some (not precise in size, nor exhaustively for representation limits) of the main net (blue and green) VW flows across regions.

Clearly, net water exporters are the Spanish regions of Andalusia, Aragon Castile-La Mancha, Castile and León and to a lesser extent La Rioja. Also in general the EU is a net importer of water while the rest of the world is clearly a net exporter. Galicia and Navarre are net exporters of green water, which is logical in a wet region where rain fed agriculture dominates, but net importers of blue water. Madrid stands out as a net importer of water, followed by Basque Country and Catalonia. Hence the heavily populated, highly industrialized regions like Madrid and Catalonia that specialize in services are net importers of goods and services and of VW. The meaning of this result is important; although these industrialized regions do not consume directly much water, their economic functioning and their inhabitants' way of life require the consumption of a high volume of water resources in other areas, generating in those areas potential bottlenecks in some dry-cyclical seasons. An increase in the population of Madrid has had an important impact on the scarce water resources from the regions of Andalucia, Murcia and Valencia, which produce many agrarian goods that sustain the population. Water resources in these regions are particularly affected during the summer season when there is scarce rainfall and tourists still require significant volumes of drinking water as well as water for recreational activities. Water is also needed to produce the energy and food that they consume. A demand model such as ours can simulate and predict such pressures through all the chains of production in each region and sector.

(Blue) Water consumptive use and Net VW export to availability ratios

One of the main insights this type of study may lead us to is to assess the pressure on water and look at ways of reducing this by understanding the uses of water resources with respect to the availability within the region. In this sense, we may observe in Map 3 how the region of Andalusia is a clear-cut net exporter of VW, being one of the regions with higher water stress.

A similar picture is obtained for Aragon or Navarre, although with less water stress problems. Also a high consumption of blue domestic water resources is found for Catalonia, the Valencian Community, Madrid, the Balearic Islands, the Canary Islands and Murcia but they are net importers of VW.

Map 3: Blue VW consumption and net exports to availability ratios

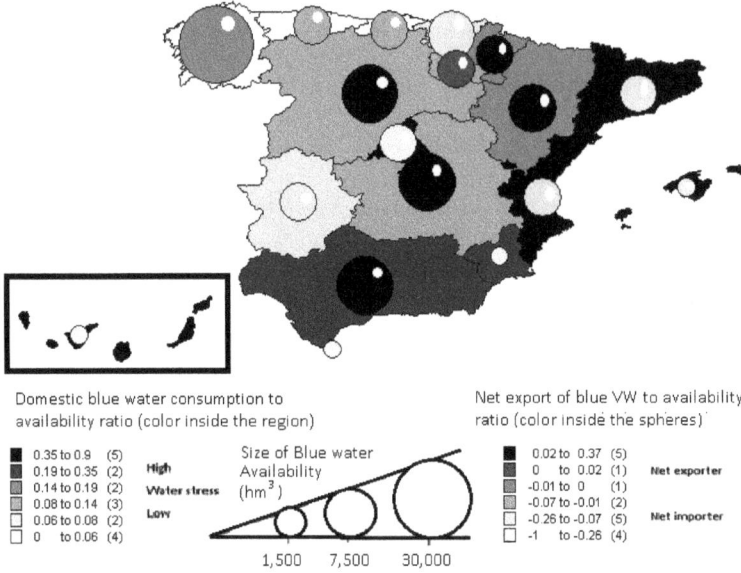

Domestic blue water consumption to availability ratio (color inside the region)

■ 0.35 to 0.9 (5)	
■ 0.19 to 0.35 (2)	**High**
■ 0.14 to 0.19 (2)	**Water stress**
□ 0.08 to 0.14 (3)	
□ 0.06 to 0.08 (2)	**Low**
□ 0 to 0.06 (4)	

Size of Blue water Availability (hm³)

1,500   7,500   30,000

Net export of blue VW to availability ratio (color inside the spheres)

■ 0.02 to 0.37 (5)	
■ 0 to 0.02 (1)	**Net exporter**
■ -0.01 to 0 (1)	
□ -0.07 to -0.01 (2)	
□ -0.26 to -0.07 (5)	**Net importer**
□ -1 to -0.26 (4)	

Note: The size of blue water availability is given in absolute terms (not put in relation to surface or population).

Conclusions and future developments

National WF in comparison with previous studies (different hypotheses)

Starting from the estimated direct domestic water consumption we account for the import and export of VW, which gives us an already well-known result: Spain is a net importer of water. The numbers are however different from those studies where the balance is estimated using national intensities, data, assumptions of trade, etc since we account for the particular regional differences which may noticeably alter the results. Looking at the different paths of the supply chain, we can also see that the Spanish demand for water comes mainly from Madrid in the final steps, which plays a significant nation-wide role in the processing and distribution of goods.

    We have also obtained the clear-cut net VW import or export character of some regions, something especially relevant when taking into account their availability of resources.

Policy implications

The study has shown some of the multiple possibilities of MRIO models to assess the issues of pressure on water or other environmental impacts. Policy implications for water management in the Spanish regions are more clear-cut with this study than with the national perspective since regional characteristics and trade are considered. In order to achieve better water use and lower the pressure on this scarce resource, it has been proposed to lower local impacts by substituting local production for water intensive imports, as estimated from the MRIO. As in the case of Andalusia, there are problematic and possibly counterintuitive situations. These are linked to issues such as intensive irrigation in areas where water resources are scarce due to climate, soil, etc. There can also be economic imperatives that override collateral damage done to resources.

Spanish regional economies are clearly interconnected and the VW flows among regions are significant. Studying these flows allows us to estimate direct and final responsibilities in the consumption of water resources achieving a more qualified estimation of the regional and basin WFs. Helping direct and indirect users to know their role in the regional production process as well as the impacts of their activities on global resources is an important step towards sustainability making MRIO models a plausible tool to provide this information.

Based on this information, new perspectives for regulatory and governance appear. For example, improved wastewater treatment frees up resources for other uses, mainly agricultural, alleviating potential pressures in the areas of origin of imported products. Similarly, efficiency in water use in the production of a good may accompany it throughout the full production chain to its final destination, generating opportunities for eco-labelling. The knowledge of impacts across the production chain may be used to determine policies such as the selection of different taxes and subsidies or recognitions based on them The recognition of quality, such as ISO certifications, protected designation of origin or geographical indication also have the potential to value-add to products that have been produced in an eco-friendly manner.

Further reading

Chapagain, A., & Hoekstra, A. (2003). Virtual water trade: a quantification of virtual water flows between nations in relation to international trade in livestock and livestock products. In A. Hoekstra (Ed.), *Virtual Water Trade: Proceedings of the International Expert Meeting on Virtual Water Trade*, (pp. 49-76). Delft, The Netherlands.

Daniels, P., Lenzen, M., & Kenway, S. (2011). The ins and outs of water use – a review of multi-region input–output analysis and water footprints for regional sustainability analysis and policy. *Economics Systems Research , 23* (4), 353-370.

Duchin, F. L. (2011). Policies and technologies for a sustainable use of water in Mexico: a scenario analysis. *Economics Systems Research , 23* (4), 387-407.

Pérez, J., Dones, M., & Llano, C. (2009). An Interregional Impact Analysis of the EU Structural Funds in Spain (1995-1999). *Papers in Regional Science , 88* (3), 509-529.

Peters, G. (2007). Opportunities and challenges for environmental MRIO modelling: Illustrations with the GTAP database. *16th International*

>   *Input-Output Conference of the International Input-Output Association
>   (IIOA).* Istanbul, Turkey: IIOA.
Wiedmann, T. (2009). A review of recent multi-region input–output models used
>   for consumption-based emission and resource accounting. *Ecological
>   Economics , 69* (1), 211-222.

Acknowledgements

The authors would like to acknowledge the directorate, and the organizers (J. Murray, A. Geschke, K. Kanemoto, D. Moran, C. Dey and M. Lenzen) of the course on Multi-Regional Input-Output Analysis at the 19th International Input-Output Conference (Alexandria, US – June 2011) for their teaching and helpful comments. We also thank very much the careful and in depth English-check carried out by Glenn Cawthorne. All the remaining errors are our sole responsibility.

Part IV: Case Studies

Still life with ginger flower
Oil on canvas 77x 91cm
Dagmar Hoffmann

Chapter 14: Case Study Using the GTAP Database: Footprint and Supply Chain Analysis of Dutch Consumption

Harry Wilting

Introduction

This chapter presents a footprint and supply-chain analysis to demonstrate the application of the GTAP database. In the footprint analysis, environmental pressures are calculated from the perspective of Dutch (private and public) consumption. The supply chain analysis provides insights into the contributions of regions and sectors all over the world to the environmental pressures arising from Dutch consumption and specific expenditures. The example used is that of the consumption of clothing in the Netherlands. Greenhouse gas (GHG) emissions and land use are the two environmental pressures that are considered.

Background

National governments, e.g. in the Netherlands, show an increasing interest in the sustainability of supply chains. This interest focuses on a need for insights into the impacts on sustainability in the producing countries and sectors that are involved in supply chains. These insights offer clues for policy-makers in designing strategies to decrease environmental and socio-economic impacts in these countries and sectors. This chapter depicts the sectors and regions in which production and related environmental pressures take place for Dutch consumption and the contributions of individual supply chains therein.

A supply chain covers all activities from suppliers, e.g. suppliers of raw materials, to the end-users of a product or service. In practice a supply chain is not a linear chain, but a complex network of interactions between companies or sectors. Many primary resources and intermediate processing steps are necessary to make a final product and also raw materials are distributed across many products and regions. The use of input-output (IO) analysis, one of the two main approaches in executing a life-cycle analysis (LCA), enables a full description and unravelling of supply chains. IO analysis offers mathematical formalisms and analytical techniques to investigate entire supply chains and the contribution from each sector (both in economic and environmental terms) to final demand in a streamlined and resource-efficient way.

Methodology and Data Sources

GHGs and land use intensities of Dutch production sectors accounting for the origin of imported products were calculated using a multi-regional input-output (MRIO) model. By combining these intensities with consumer demand in the Netherlands, the total consumption-related environmental pressures (footprint) were calculated. Direct GHG emissions and land use by Dutch consumers, such as residential emissions from heating or car use are not covered by an IO analysis and were therefore added separately to the calculated 'indirect' pressures that take place in production sectors. By using the MRIO formalism, imports and region-specific production technologies are accounted for. The MRIO formalism also enables the investigation of the regions and sectors in which environmental pressures arising from Dutch consumption mainly take place (supply chain analysis).

The MRIO model was built by combining economic and environmental data. The economic data consist of national IO tables and all bilateral trade flows, and were derived from the GTAP database (version 6), describing the global economy in 2001 (see Chapter 3) This version of the GTAP database contains input-output data for 87 regions and 57 sectors. For the MRIO model used in this chapter the 87 regions were aggregated to 13 regions, viz. 12 aggregated world regions and the Netherlands. Data on GHG emissions and land use were obtained from international environmental databases. The GHGs that were included are carbon dioxide (CO_2), methane (CH_4), nitrous oxide (N_2O) and a group of fluorinated (F) gases. Land use data concerned land use for crops, pasture, forestry products and built-up land, which covers urban land and land for infrastructure. All land-use data apply to physical areas and no distinction was made between extensive and intensive use of the land. Especially for pastureland, there are huge differences in local land-use intensities between countries, e.g. the number of hectares to produce a million dollars worth of milk.

Figure 1: Overview of region-sector combinations in total consumption and individual consumption items (S = Sector, D = Direct consumption, T = Total).

Footprint indicator	Region 1 S1	S2	S3	⋮	Sn	D	Region 2 S1	S2	S3	⋮	Sn	D	…..	Region m S1	S2	S3	⋮	Sn	D	T
Consumption category 1																				
Consumption category 2																				
Consumption category 3																				
…..																				
Consumption category k																				
Total																				

Figure 1 gives a schematic overview of the outcomes of the calculations for a specific footprint indicator (e.g. GHG emissions). The rows depict the pressures in region-sector combinations for individual consumption items (or supply chains). The row totals depict the overall pressure of a complete supply chain. The sum of the row totals (depicted by the cell in the lower right corner) is the total environmental pressure for the specific footprint indicator from the perspective of consumption of the country. The column totals depict the overall environmental pressure in a certain region-sector combination for total Dutch consumption (summing up over consumption items).

Outcomes

Total Dutch GHG and land footprint

Before discussing the contributions of specific sectors and regions in the Dutch footprint, the overall outcomes in terms of the environmental pressures from the consumption perspective are presented (corresponding to the final column in Figure 1). GHG emissions and land use are considered from the perspective of individual consumption items (here aggregated to nine categories), in which the direct environmental pressures in the consumption stage are included.

Figure 2: Contributions in expenditures and related environmental pressures for nine aggregated consumption categories.

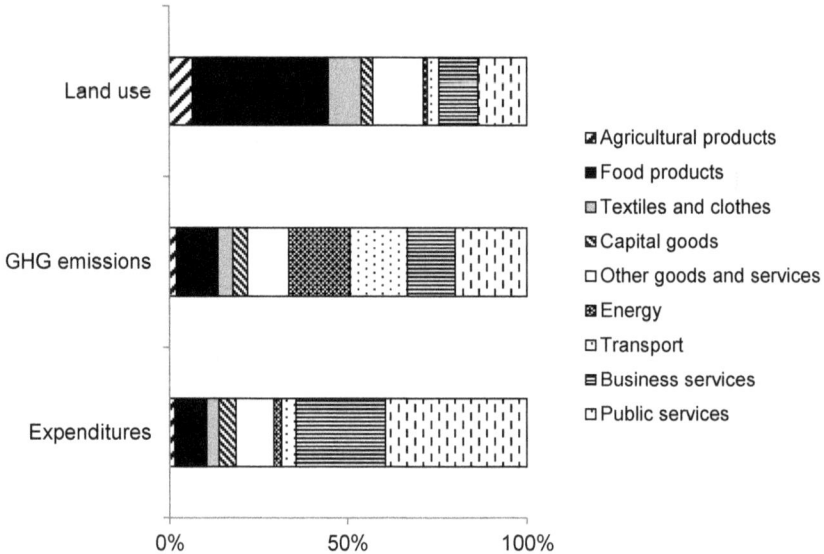

Figure 2 shows for Dutch consumption of goods and services the expenditures and the corresponding supply-chain GHG emissions and land use. Expenditures on financial and business services, public administration and other services make an important contribution to Dutch consumption patterns (private and public). Expenditures with high GHG emissions are services and of course oil products (in transport), natural gas (which includes emissions during consumption) and electricity. Expenditures associated with high land use are food products (like dairy products and meat) and other wood and paper products (in other goods and services). Expenditures with high pressures have a high share in consumption, such as services, or have high intensities. Expenditures with high GHG intensities are energy products and transport. Land use intensities are high for forestry products and products from cattle (meat and leather).

Specific chain: clothing

This section discusses the unravelling of a specific supply chain according to consumption in regions and sectors in which pressures take place (this corresponds to a row in Figure 1). As an example, the outcomes for the clothing chain are discussed. The clothing chain is an interesting supply chain for further investigation, since not all clothes are produced in a sustainable way. There are environmental pressures in producing raw materials for the clothing industry. Cotton cultivation requires pesticides, fertilizers and water, but other natural and synthetic resources have an environmental impact too. Furthermore, the processing of raw resources in the textile industry uses chemical products like chlorine or dye that may result in pollution. In the use and disposal stage of the chain environmental pressure may occur, for example in cleaning or washing. In some countries, labour circumstances of needlewomen are bad due to low wages, long working weeks and monotonous work. Furthermore, a small percentage of child labour exists in the clothing industry.

Figure 3: Greenhouse gas intensities for the production of clothes per region (kg CO_2-eq./US $); for each intensity, it is indicated in which region emissions take place.

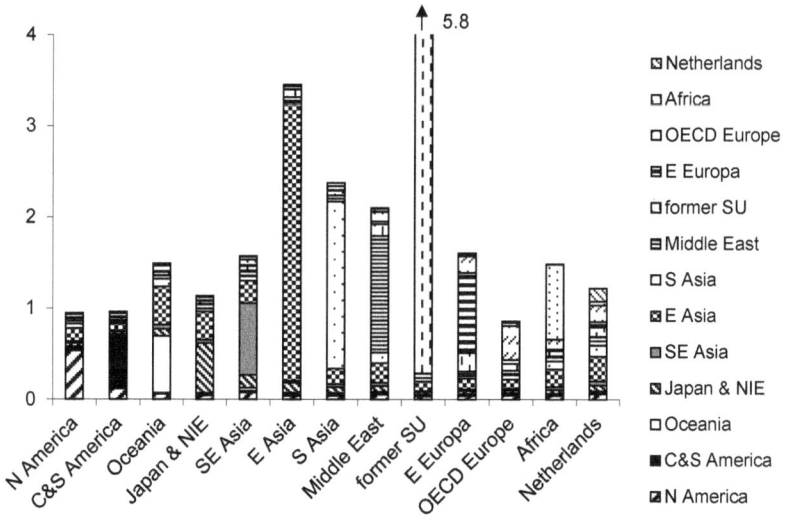

The supply chain of clothing differs per region, e.g. clothes produced in North America have a different supply chain than clothes produced in China. Figure 3 shows per region the differences in GHG emissions per unit of production over the whole supply chain. The GHG intensities of production of clothes show the highest values for former Soviet Union and the China region (E Asia) due to high emissions in the electricity sector in these regions. Please note that due to price differences between regions, a dollar of clothes in one region represents a higher quantity of clothes than a dollar of clothes in another region. So, the differences in GHG intensities are the combined effect of differences in efficiencies and prices.

Figure 4: Contributions of regions to Dutch consumption of clothes showing related value added and environmental pressures.

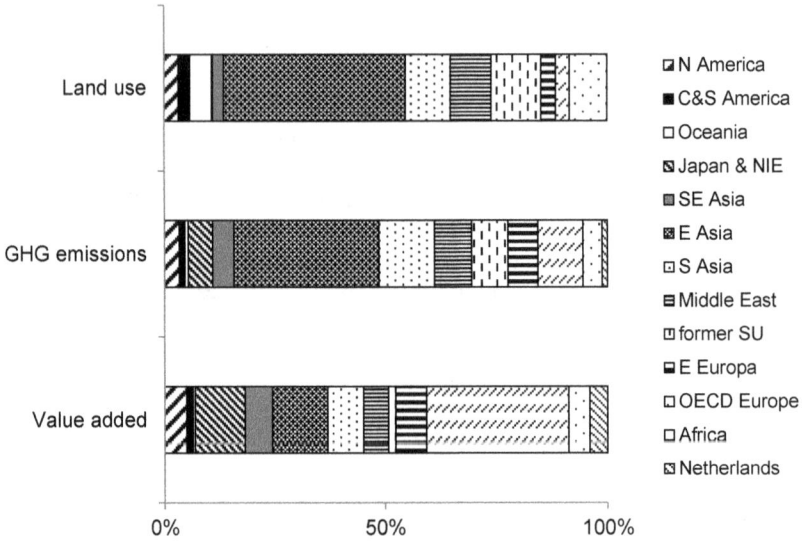

The main part of the clothing supply chain leading to Dutch consumption of clothes lies outside the Netherlands. Clothes are imported from several regions and the production of textiles for the Dutch clothing industry mainly takes place abroad. Almost 50% of clothes bought by Dutch consumers are of European origin, but this includes the upgrading of clothes or textiles imported from China or other Asian countries. More than 40% of clothes consumed in the Netherlands originate directly from Asian countries. The lower bar in Figure 4 gives for Dutch consumption of clothes an overview of the regions where value added is created. Please note that in this case study, the chain ends at the final production stage, i.e. the clothing industry. Therefore, the trade margins produced in wholesale and retail, which mainly concerns the Netherlands, are not included in Figures 4 and 5. The upper bars in Figure 4 show where environmental pressures take place. For GHG emissions and land use the situation is quite different from value added. More than 50% of environmental pressure arising from the clothes chain takes place in Asia.

Figure 5: Contributions of sectors to value added and environmental pressures arising from Dutch consumption of clothes.

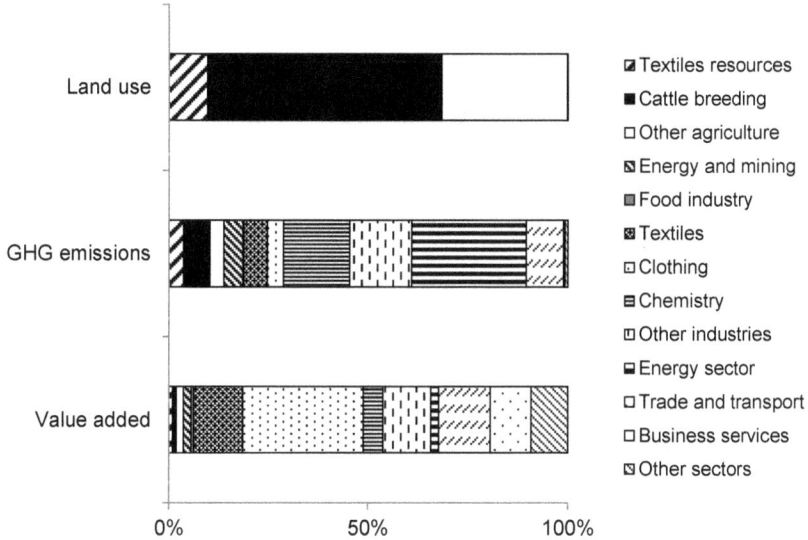

Figure 5 shows the economic and environmental effects over sectors. Sectors directly related to clothes (textiles and clothing) have a high share in value added, but only have a small share in GHG emissions. Sectors with a high share of GHG emissions are the energy sector, chemistry, and other manufacturing. Cattle breeding and other agriculture have a high contribution in land use in relation to textile resources. The high share of cattle breeding is caused by the use of local land-use intensities in the model (as mentioned above).

Unravelling of the Dutch footprint

Similar analyses as those carried out for the clothing chain can be done for all consumption categories. This results in an overview of Dutch consumption and related pressures in region-sector combinations. Summing up over all consumption categories leads to the sectoral and regional disaggregation of environmental pressures arising from Dutch consumption (the last row in Figure 1). Using the MRIO model it is not necessary to first calculate the outcomes for all individual consumption items. The IO formalism enables the calculation of pressures of total consumption per sector-region combination directly.

About 90% of the goods and services consumed by Dutch government and households are of Dutch origin, which means that the final producer is located in the Netherlands. Since Dutch producers in their turn use imports, the percentage of value added created by Dutch production as a share of total value added for Dutch consumption is less than 90% (Figure 6). Nevertheless, value added creation arising from Dutch consumption mainly takes place in the Netherlands and surrounding countries. In 2001, more than 80% of value added was created in OECD Europe (including the Netherlands).

Figure 6: Contributions per region in value added and environmental pressures arising from Dutch consumption.

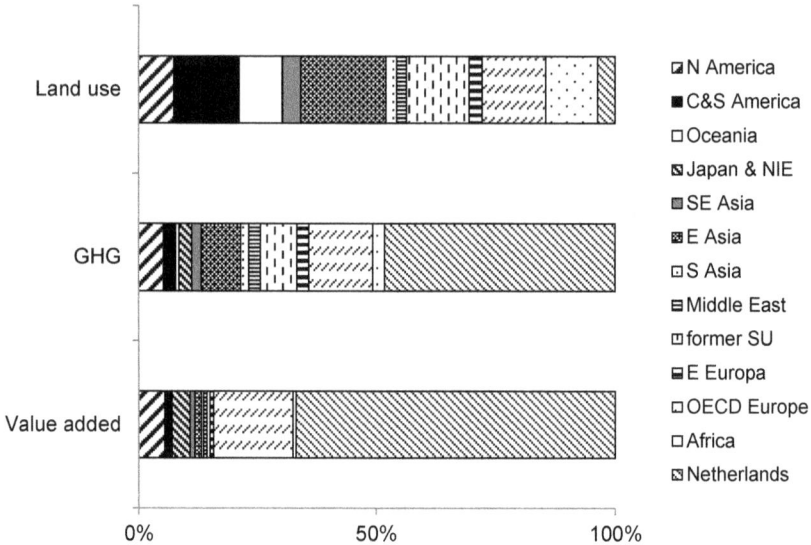

Contrary to value added creation, which usually occurs at the end of chains, environmental pressures occur mainly in the beginning of chains, for example at resource extraction and processing. More than 50% of GHG emissions arising from Dutch consumption take place abroad. These gases are emitted especially in the OECD Europe region, but the regions of China, the former Soviet Union and North America have substantial contributions too. In land use, the situation is even more extreme; more than 90% of land use is outside the borders of the Netherlands. It is noticed again, that the calculation is based on local (physical) hectares with very low efficiencies for the regions South-America, China and Oceania. An important part of the land-related production in the Netherlands is exported (mainly to other European countries), and is therefore not assigned to Dutch consumption.

In each region, value added and pressures arising from Dutch consumption are distributed over the separate sectors. Sectors with high shares in total value added arising from Dutch consumption are public and business services in the Netherlands. The sector other business services in OECD Europe is the highest non-Dutch sector in the list of sectors contributing to value added created by Dutch consumption with a contribution of 2.5%. Table 1 shows the region-sector combinations with the highest contributions in GHG emissions and land use. Direct GHG emissions of consumers contribute almost 20% in total GHG emissions arising from Dutch consumption. Furthermore, the production of electricity is a main contributor to GHG emissions arising from Dutch consumption; the electricity sectors in the regions Netherlands, OECD Europe, China, former Soviet Union and North America, all have a high placing in the top ten. Electricity use is essential in many production processes. Other region-sector combinations with high contributions in GHG emissions are more diverse. The land-related sectors agriculture and forestry highly contribute to the land use arising from Dutch consumption. Agricultural sectors with high contributions are the cattle and dairy sectors and the production of oil seeds (for feed).

Table 1: Region-sector combinations with the highest contributions in GHG emissions and land use for Dutch consumption with shares in percentages.

Greenhouse gases			Land use		
Region	Sector	%	Region	Sector	%
Netherlands	direct	18.5	E Asia	cattle	13.4
Netherlands	electricity	12.3	OECD Europe	forestry	5.5
Netherlands	public services	3.1	C&S America	cattle	5.3
OECD Europe	electricity	3.0	Oceania	cattle	4.1
E Asia	electricity	2.2	C&S America	oil seeds	4.1
former SU	petroleum and coal	2.1	Oceania	raw milk	3.9
former SU	electricity	2.1	Africa	other crops	3.8
Netherlands	land transport	1.7	N America	oil seeds	3.2
OECD Europe	gas	1.6	former SU	forestry	3.2
N America	electricity	1.5	Africa	cattle	3.2
OECD Europe	petroleum and coal	1.5	former SU	cattle	3.0
E Asia	minerals	1.4	former SU	raw milk	2.7
Netherlands	air transport	1.3	OECD Europe	wheat	2.5
Netherlands	public services	1.2	OECD Europe	other grains	2.2
Netherlands	raw milk	1.1	N America	forestry	2.1
OECD Europe	land transport	1.1	SE Asia	forestry	1.8
Netherlands	trade	1.1	former SU	wheat	1.6
Netherlands	chemical products	1.0	Africa	raw milk	1.5
OECD Europe	minerals	1.0	Netherlands	direct	1.4
OECD Europe	chemical products	0.9	Netherlands	raw milk	1.3

Final remarks

The analysis presented in this chapter offers insights, both for researchers and for policy makers, about the sectors and regions where production and related pressures take place for Dutch consumption. The outcome of the analysis is a detailed picture consisting of region-sector combinations with high contributions to the pressure of total Dutch consumption or specific consumption items. The analysis was carried out for GHG and land use only, since the required data were easily accessible. It is imaginable that the analysis can be extended to other environmental pressures without much effort in searching out the relevant data. In a complete sustainability analysis, social and economic aspects have to be considered too. In most cases, data on social pressures are not available on a global level for all regions and sectors. However in some cases, specific data on e.g. child labour or bad working conditions are available for specific region-sector combinations. The MRIO model can be used to determine which part of production in that specific sector and region is intended for Dutch consumption. This might be an indication of the social pressure in that region arising from Dutch consumption. However, in general this share will be small; most sectors produce mainly for their own domestic demand.

The analysis of individual consumption items offers clues for selecting chains for further research, e.g. concerning the identification of options directed at reducing pressures. Selection criteria may be the share in pressures arising from Dutch consumption and the pressures per unit of consumption. So, interesting chains for further investigation are chains with high absolute pressures or chains with high relative pressures (pressure per unit) compared to other chains. The insights in the pressures along supply chains may be useful for optimising supply chains over country borders. In order to produce, e.g. clothes with fewer pressures, insights are required concerning in which region-sector combinations pressures can be reduced. Such an optimisation, which requires an international approach, may involve a shift of foreign production to more efficient production in the Netherlands or an improvement to technology elsewhere. However, the optimisation has to be carried out with care, since an optimisation for greenhouse gases may lead to less optimal outcomes for other environmental themes, like acidification or land use. These themes are specific for certain regions, unlike climate change, which is a global problem. Furthermore, optimising on ecological aspects alone is too narrow from the viewpoint of sustainability. Therefore, in optimising over chains, economic and social aspects have to be considered too.

The supply chain analysis offers insights for policy makers, e.g. that a substantial part of the pressures arising from Dutch consumption occurs outside the borders of the Netherlands. Dutch policy makers may use instruments directed at the domestic actors and at actors abroad in order to reduce these pressures. Instruments for Dutch policy directed at actors abroad are the Kyoto mechanisms (Joint Implementation and Clean Development Mechanism). These mechanisms concern the financing of projects in other countries in order to reduce the emissions in those countries. The financing of new technologies in region-sector combinations with high pressures for Dutch consumption will decrease the Dutch footprint abroad. Furthermore, policy makers might reduce foreign emissions via Dutch producers and consumers. Producers can request of their suppliers stringent conditions concerning the sustainability of their production processes. Furthermore, financial instruments or hallmarks concerning the sustainability of supply chains may be used to stimulate consumers in choosing more sustainable consumption patterns.

Further reading

MNP. (2008). The Netherlands in a Sustainable World: Poverty, Climate and Biodiversity; Second Sustainability Outlook. Bilthoven, The Netherlands: Netherlands Environmental Assessment Agency.

Wiedmann, T., & Lenzen, M. (2007). Unravelling the Impacts of Supply Chains – A New Triple-Bottom-Line Accounting Approach. Centre for Sustainability Accounting. York: Centre for Sustainability Accounting.

Wilting, H., & Vringer, K. (2009). Carbon and Land Use Accounting from a Producer's and a Consumer's Perspective; an Empirical Examination covering the World. *Economics Systems Research*, 21, 291-310.

Acknowledgements

The author would like to acknowledge Maaike Wilting for her comments on an earlier draft of this chapter.

Chapter 15: Consumption-Based Inventory of Global Land Use: An Application of the GTAP Database

Yang Yu, Kuishuang Feng, and Klaus Hubacek

Introduction

Anthropogenic land use has increased rapidly over the past few decades mainly as a result of population growth, increase of economic activities, urbanization and changes in lifestyles. Globally, approximately 29% of the land surface has been converted to agricultural and urban or built-up areas. It is projected that an additional 33% of global land could be converted over the next 100 years (WRI, 2000). In a globalized world, goods and services derived from land use that are consumed in one country are often produced in another country and exchanged via international trade. Local land use changes are increasingly triggered by demands for products that are consumed elsewhere and that are part of global supply chains oftentimes involving large geographical distances. Wealthier countries consume a larger amount of goods and services from both domestic and international markets, and thus impose pressure not only on their domestic land resources, but also appropriate land in other countries. This is most evident with regards to land for food production. Wealthier countries with high consumption and scarce land resources appropriate farmland from foreign countries either through leasing farmland in other countries (sometimes referred to as 'land grab') or through importing food products from other countries (a more subtle land grab) (Görgen et al., 2009). It increases the concern about land as a scarce resource, but also leads to new competition with local users and their livelihood needs (Cecilie and Anette, 2010). With the increase in demand for land-intensive products globally, such as food, fiber and biofuel, it is important to understand how land is used in the production of goods and services to meet future demand. Thus we need to track land use along global supply chains to examine how much and for what purpose one country has been acquiring land in other countries' territories in order to support consumption and lifestyles of their own citizens. As explained in Chapter 1, multiregional input-output (MRIO) analysis is a powerful tool to analyze global supply chains. By using a global MRIO model, we calculate domestic and foreign land appropriation for 113 global countries/regions defined in the Global Trade Analysis Project (GTAP).

Results

In this study, domestic land appropriation for a country is the domestic land used for producing goods and services for domestic consumption; foreign land appropriation is the foreign land used for producing goods and services for domestic consumption. Figure 1 shows domestic and foreign land appropriation for 15 selected countries across the world. From the figure we see that the US, the world's largest land consumer, appropriates 31% of its land from other countries to meet its own consumption. The share of foreign land in total land appropriation for Japan, Germany and the UK is close to 90%. With limited land resources, these countries appropriate a large amount of foreign land through importing goods to satisfy their own needs. As emerging economies with vast populations, China appropriates about 14% and India appropriates 12% of their total land use from foreign countries. In contrast, Brazil, Russia, Australia and Canada with abundant land resources and very low population density produce their goods for their own final consumption mostly from their domestic land. Figure 1 also shows that African countries, such as Tanzania and Zambia, appropriate most land for their domestic consumption from their own territory, the same as developing countries in Asia such as Indonesia.

Figure 1: Domestic and foreign land appropriation for fifteen selected countries across the world. The fifteen countries are selected from different continents across the globe with different socio-economic and geographical characteristics.

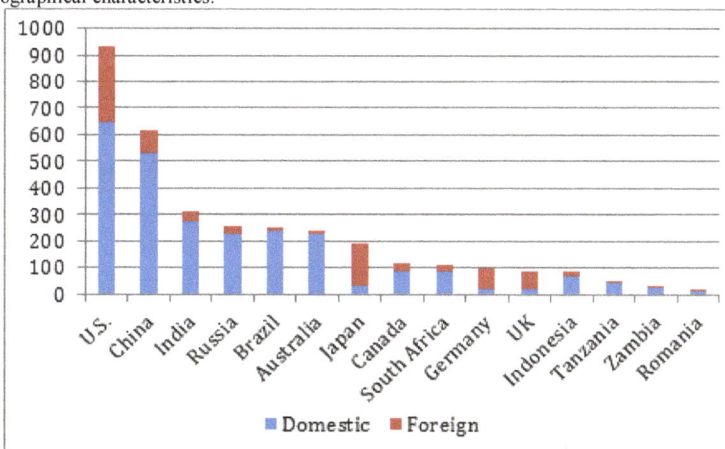

In addition to the total foreign land appropriation, it is also important to know in which country the foreign land is appropriated. In general, countries with relatively large areas of land tend to produce and export more land-intensive products, while countries with insufficient land resources tend to import land intensive products from other countries. It can be seen from Figure 2 that countries with vast amounts of land such as Russia, Canada, Australia and Brazil, have a negative net balance and 'export' their land to other regions (or displace land for other uses) through exporting land intensive products. In most of these countries the appropriated land by foreign countries through exports was more than ten times larger than the foreign land appropriation through imports. However, United States as a land-rich country is also the world's largest land consumer and appropriates nearly 300 Mha land from other countries.

Figure 2: Land appropriation through imports and exports for the twelve largest global land consumers (Mha).

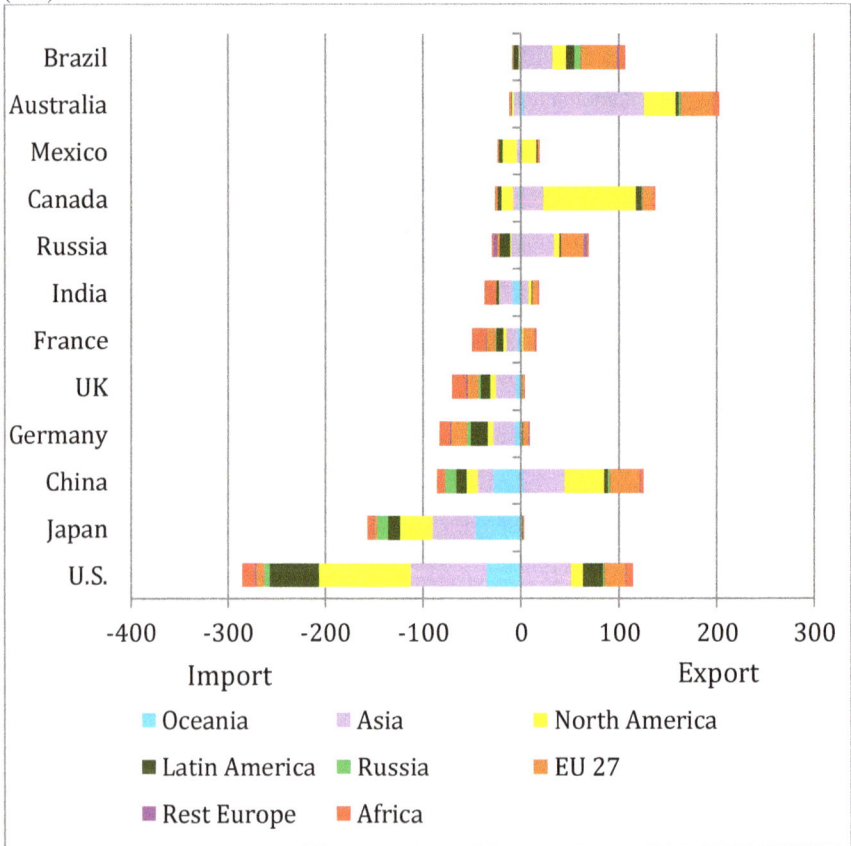

Figure 2 shows land appropriation through imports and exports of the top 12 land consuming countries. From the figure we can see that the US has the largest foreign land appropriation. About 40% of the US total foreign land is appropriated from its neighbours in North America such as Canada. The US also uses a large amount of land from other world regions, such as Latin America (21% of US foreign land appropriation), Asia (16%), Russia (8%) and Africa (5%). Japan is the world's second largest foreign land consumer, appropriating land from Russia (30% of Japan's foreign land appropriation), Asia (20%), Oceania (19%) and North America (18%). China, as an emerging economy, is ranked as the third largest foreign land consumer, using land mainly from Russia and Africa. EU countries have a substantial proportion (between 13% and 34%) of their foreign land appropriation from Africa.

EU countries also require a large amount of land from Latin America, in particular from Brazil. Thus consumption in EU countries is linked to deforestation in the Brazilian Amazon through consumption of agricultural products produced from Brazil, which triggers the land conversion from forestland to agricultural land, and through consumption of substantial amounts of wood products and inputs of wood products in production processes.

Overall, global land appropriation by the US, EU and emerging economies such as China and India may not only put pressure on local food security in particular in Africa, but also cause deforestation in the Amazon, Canada and Russia which may further impact on global climate change and damage local ecosystems.

Further readings

Cecilie, F., & Anette, R. (2010). *Land grab in Africa: Emerging land system drivers in a teleconnected world. Series title,* The Global Land Project. Copenhagen: The Global Land Project.

Görgen, M., Rudloff, B., Simons, J., Üllenberg, A., Väth, S., & Wimmer, L. (2009). *Foreign direct investment (FDI) in land in developing countries.* Tyskland: Federal Ministry for Economic Cooperation and Development Division 45.

WRI. (2000). *A guide to world resources 2000 - 2001: People and ecosystems: The fraying web of life,.* Washington, DC: World Resources Institute.

Chapter 16: Accounting for Environmental Responsibility: A Case of Asian Countries Using the Asian International Input-Output Model

Xin Zhou and Alexandra Marques

Introduction

The main driver for climate change is the accumulation of greenhouse gases (GHG) in the atmosphere, which results from anthropogenic activities, namely the burning of fossil fuels and land use changes. The climate system is a common resource; its protection requires joint efforts and global collaboration. As the first milestone in tackling climate change, the Kyoto Protocol committed 37 industrialized countries and the European Union to collectively reduce their GHG emissions by an average of 5% against 1990 levels.

Due to the 'common but differentiated responsibilities', developing countries did not commit themselves to any quantified mitigation. Though they account for less than one quarter of historical emissions, over three quarters of future emissions growth will likely come from today's developing countries because of their rapid population and GDP growth. Therefore all emitting nations should take some responsibility.

In order to determine the emissions each country is responsible for and to monitor the progress towards established targets, Kyoto requires that countries report, through national GHG inventories, "emissions and removals taking place within national (including administered) territories and offshore areas over which the country has jurisdiction"[1]. Through these national GHG inventories only direct emissions are accounted for, international transportation and indirect effects associated with international trade are excluded. This is the ***producer responsibility principle***.

There are both pros and cons for the producer responsibility principle. On the one hand, producer responsibility is underpinned by the polluter-pays-principle, which has been endorsed by the OECD countries since mid-1970s. In practice, direct emissions are easier to estimate, monitor and report. Also accounting for emissions within the boundary of national jurisdiction respects the sovereignty of states.

On the other hand, the shortcomings of this approach are numerous and their consequences are important. The producer responsibility principle makes impossible the allocation of international transportation or other indirect effects, allows for carbon leakage phenomena through international trade and creates issues of fairness and competitiveness difficult to overcome[2,3] Kyoto places all responsibility on producers. A country whose economy is mainly supported by exports will have, comparatively, more direct emissions than a non-exporting country. This framework is unlikely to be accepted by rapidly developing countries, like China or India, whose economies are highly dependent on exports and have the highest CO_2 emissions growth rates[4].

To address these shortcomings, it is necessary to incorporate international trade and consider other responsibility principles in assessing national inventories. 'Embodied emissions' is such an indicator which tries to address consumer responsibility by assessing the emissions generated from all upstream stages, no matter from where, in the supply chain. In contrast to the producer responsibility principle, the ***consumer responsibility principle*** requires consumers to be responsible for all upstream emissions embodied in their consumption.

As a counterpart to the upstream responsibility in a supply chain, the downstream responsibility in a sales chain requires suppliers be responsible for the emissions generated from all downstream stages. Because of their supply, the downstream producers are enabled directly or indirectly to produce and hence to emit. The suppliers benefit from the emissions by obtaining income and therefore should assume the responsibility. 'Enabled emissions' is used as an indicator to assess such ***downstream responsibility*** or ***income responsibility principle*** (see Marques *et al.*, 2012 in Further readings).

---

[1] UNFCCC (1998). *Kyoto Protocol to the United Nations Framework Convention on Climate Change*. Technical Report. Bonn: United Nations Framework Convention on Climate Change (UNFCCC).

[2] Peters, G., & Hertwich, E. (2008). CO_2 embodied in international trade with implications for global climate policy. *Environmental Science and Technology* 42, 1401-1407.

[3] Whalley, J., & Walsh, S. (2009). Bringing the Copenhagen global climate change negotiations to conclusion. *CESifo Economic Studies* 55, 255-285.

[4] Raupach, M.R., Marland, G., Ciais, P., Quéré, C.L., Canadell, J.G., Klepper G., & Field, C.B. (2007). *Global and regional drivers of accelerating CO_2 emissions*. Proceedings of the National Academy of Sciences, 485, 10288-10293.

Input-output analysis can be used to allocate upstream emissions or downstream emissions systematically while avoiding the double-counting problem. In doing so, ultimate upstream responsibilities are allocated to the consumers of the final products (such as households) by using the Leontief inverse and ultimate downstream responsibilities are allocated to the providers of primary factors (such as workers, investors, land owners, etc.) by using the Ghosh inverse.

By either of the approaches, however, the nature of each agent in the production chain being both a supplier and a customer at the same time cannot be captured, leaving the fairness issue unsolved. New metrics, which aim to recognise shared producer and consumer responsibility (Gallego and Lenzen, 2005; Lenzen *et al.*, 2007) or shared upstream and downstream responsibility (Rodrigues and Domingos, 2008), are therefore introduced.

In this chapter we illustrate an application of the Asian International Input-Output Model (AIO) in assigning environmental responsibility across countries and compare national inventories based on producer responsibility, consumer responsibility, income responsibility and shared environmental responsibility.

Asian International Input-Output Model

AIO was constructed by the Institute of Developing Economies, Japan External Trade Organisation (IDE-JETRO)[5]. IDE developed the first AIO for 1985 and updated it every five years until the year 2000. The AIO 2000 is compiled for 7 sectors, 24 sectors and 76 sectors based on different sectoral aggregation schemes. In our calculations we apply the 24-sector version of the AIO 2000. As introduced in Chapter 4, AIO 2000 is a Chenery-Moses type of multiregional model established based on national IO tables of ten economies including Indonesia, Malaysia, the Philippines, Singapore, Thailand, Mainland China, Taiwan, the Republic of Korea, Japan and the USA.

The simplified framework of the AIO 2000 is shown in Figure 1. Matrix AX represents interregional and inter-industrial transactions of intermediate goods. Matrix F represents final demand in ten economies that are supplied by themselves. E represents exports from ten economies to Hong Kong, EU and the rest of the world (ROW), respectively. Hong Kong and EU are treated separated from the rest of the world to recognize them as important trading partners of the ten economies. X represents total outputs by row or total inputs by column of ten regions. AX, F, E and X are expressed in producer prices. International transportation costs for trade among ten economies are presented as BA for intermediate goods and BF for final goods. Imports from Hong Kong, EU and ROW to ten economies are represented as import matrix IA for intermediate goods and IF for final goods in CIF (cost, insurance and freight) prices. DA and DF represent duties and taxes for all interregional trade and imports from Hong Kong, EU and ROW. V is value added, further disaggregated into wages and

---

[5] IDE-JETRO (2006). Asian International Input-Output Table 2000, Vol. 1, Explanatory Notes. Chiba: Institute of Developing Economies, Japan External Trade Organization (IDE-JETRO).

salary, operating surplus, depreciation of fixed capital and indirect taxes less subsidies.

Figure 1: Simplified framework of AIO 2000.

	Intermediate demand in ten economies	Final demand in ten economies	Export to Hong Kong, EU and ROW	Total outputs
Supply from ten economies	AX	F	E	X
Freight & insurance	BA	BF		
Imports from HK, EU and ROW	IA	IF		
Duties & taxes	DA	DF		
Value added	V			
Total inputs	X			

Source: IDE-JETRO (2006)

Reading by row, total outputs are distributed to satisfy intermediate demand, final demand and exports. By column, total inputs include purchases of intermediate goods and imports for production, payment for international transportation and tariff and payment for primary factors, such as labor, capital and governmental taxes, etc.

Accounting for National Emissions Based on Different Responsibility Indicators

We use an example of automobiles manufactured in Japan (see Box 1) to illustrate the ways of accounting for national emissions based on different responsibility indicators.

Box 1 Manufacture of automobiles

Automobile manufacture in Japan is completed through international purchase of minerals, components and parts that are produced in different countries around the world. Iron ore is imported from Australia for the production of iron and steel that is used for the car body. Electronic components are produced either domestically or imported from Thailand and China. Non-metallic ores are imported from Australia or Vietnam to produce automobile glass. Rubber is imported from South-East Asian countries for the production of tires. Plastic parts and textiles for doors, seat covers and car interior, etc. are manufactured by petrochemical industries which import naphtha and oil from Middle East countries. All the components and parts are then sent to the automobile assembly industry to produce a finished car for sale.

Emissions from different stages of production

Emissions data and source countries shown below are used only for illustrative purpose and do not represent the real situation.

Stages/actors	Direct emissions	Source countries
Iron ore extraction	0.3 unit	Australia
Iron/steel and car body processing	19 units	Japan
Electronic components manufacturing	2 units	Thailand (1/2) and China (1/2)
Non-metallic mineral extraction	0.3 unit	Australia
Automobile glass processing	4.3 unit	Japan
Rubber production	1 unit	Indonesia
Tire manufacturing	0.5 unit	Japan
Oil extraction	2 units	Middle East Countries
Plastic parts and textile manufacture	1 unit	Japan
Electricity generation	26 units	Japan
International transportation	33 units	Japan
Automobile assembly	1.5 units	Japan
Total	90.9 units	

Figure 2: The upstream carbon responsibility, passed on from upstream suppliers to the final consumer through a simplified supply chain of automobile production (u$ = dollar units).

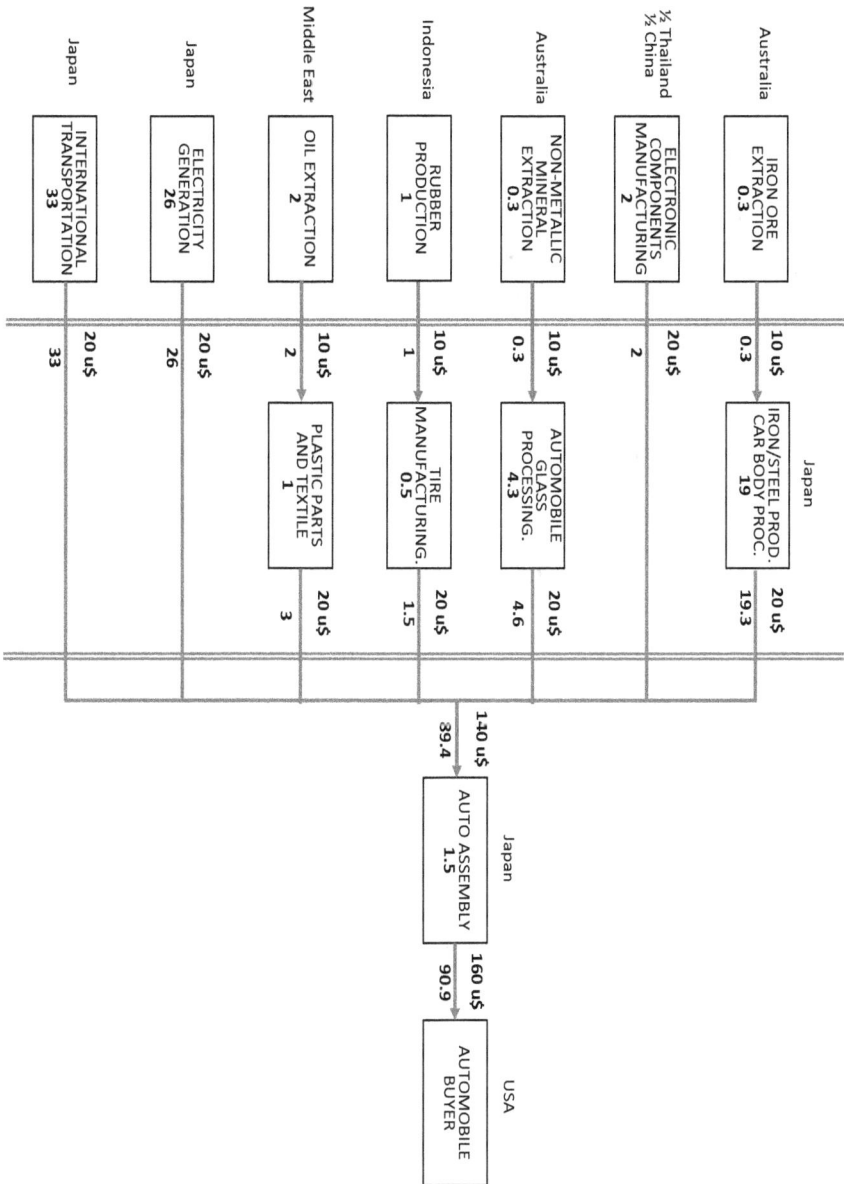

Japan — INTERNATIONAL TRANSPORTATION 33 — 33 — 20 u$

Japan — ELECTRICITY GENERATION 26 — 26 — 20 u$

Middle East — OIL EXTRACTION 2 — 2 — 10 u$ — PLASTIC PARTS AND TEXTILE 1 — 3 — 20 u$

Indonesia — RUBBER PRODUCTION 1 — 1 — 10 u$ — TIRE MANUFACTURING 0.5 — 1.5 — 20 u$

Australia — NON-METALLIC MINERAL EXTRACTION 0.3 — 0.3 — 10 u$ — AUTOMOBILE GLASS PROCESSING 4.3 — 4.6 — 20 u$

½ Thailand ½ China — ELECTRONIC COMPONENTS MANUFACTURING 2 — 2 — 20 u$

Australia — IRON ORE EXTRACTION 0.3 — 0.3 — 10 u$

Japan — IRON/STEEL PROD. CAR BODY PROC. 19 — 19.3 — 20 u$

Japan — AUTO ASSEMBLY 1.5 — 140 u$ 89.4 — 160 u$ 90.9

USA — AUTOMOBILE BUYER

By producer responsibility indicator, the source country is responsible for the corresponding direct emissions. Therefore for one automobile manufactured in Japan, Japan is responsible for 19+4.3+0.5+1+26+33+1.5=85.3 units of emissions; Australia is responsible for 0.3+0.3=0.6 unit; Thailand and China are responsible for 1 unit, respectively; Indonesia is responsible for 1 unit, and Middle East countries are responsible for 2 units of emissions.

By consumer responsibility indicator, the person who bought the car, i.e. the final consumer, is responsible for all the emissions generated from upstream productions, i.e. 90.9 units. If the person resides in the USA, then the USA, and none of the emissions source countries, is responsible for the emissions. The concept of consumer responsibility within this example is schematized in Fig.2.

Under the income responsibility principle, the providers of primary factors who enable emissions downstream assume the responsibility. Each sector employs primary factors that receive a certain amount of income and therefore the providers of the primary factors hold part of income responsibility. Figure 3 presents the downstream carbon responsibility that flows from the final consumer to the providers of primary factors of production. Under this approach, each sector in the supply chain will share its direct and downstream emissions with its suppliers (both direct and indirect) based on their income. We also assume that the providers of primary factors reside in the same country as the industry that they provide for.

For the example in Box 1, the emissions generated by the auto assembly manufacturer (1.5 units) are allocated among the providers of primary factors both for the industry itself and for seven upstream intermediate suppliers (the iron and steel production and car body processing, electronic components manufacturing, and automobile glass processing, etc.). Since in this case the providers of primary factors for the auto assembly received 20 u\$ and the auto assembly paid each of the seven upstream suppliers the same 20 u\$, the income responsibility of the primary factors of the auto assembly productions is 1.5 units shared equally between the seven upstream suppliers and the auto assembly industry itself: $1.5/8 = 0.19$ units of emissions. It is the same amount of emissions that the auto-assembly passes to each of its direct upstream suppliers. These upstream suppliers will then transfer their income responsibility (direct emissions plus the emissions they enabled in the auto assembly) to the providers of primary factors for the industry itself and for other upstream suppliers.

Following this rationale, we see that the providers of primary factors of the iron ore extraction sector enabled the emissions of 9.895 units (i.e. 9.595+0.3) and the providers of the primary factors of the non-metallic extraction sector 2.545 units. Thus, in this automobile production chain, Australia enabled the emissions of 12.44 units, in order to receive 20 u\$ of economic benefit. Electronic components manufacturers' primary factor providers enabled the emissions of 2.19 units, of which 1.095 units were enabled by Thailand and 1.095 units enabled by China; each receiving an economic benefit of 10 u\$. The providers of primary factors of the rubber production sector in Indonesia enabled the generation of 1.345 units of emissions and the providers of the primary factors of the oil extraction sector in the Middle East enabled 2.595 units of emissions. In this supply chain, Japan is the country whose providers of primary factors enabled the most emissions, and the country that received the most economic benefit as well. The income responsibility of Japan is the sum of the emissions enabled by the providers of the primary factors for electricity generation, international transportation, iron and steel production and car body processing, automobile car glass processing, tire manufacturing, plastic parts and textile manufacturing and the automobile assembly sectors. This corresponds to

33.19+26.19+0.595+0.345+2.245 +9.595+0.19=72.35 units of emissions and an economic benefit of 100 u$. The USA, where the consumer resides, is not responsible for any emissions.

Figure 3: The downstream carbon responsibility, passed on from downstream customers to upstream suppliers in a simplified supply chain of automobile production.

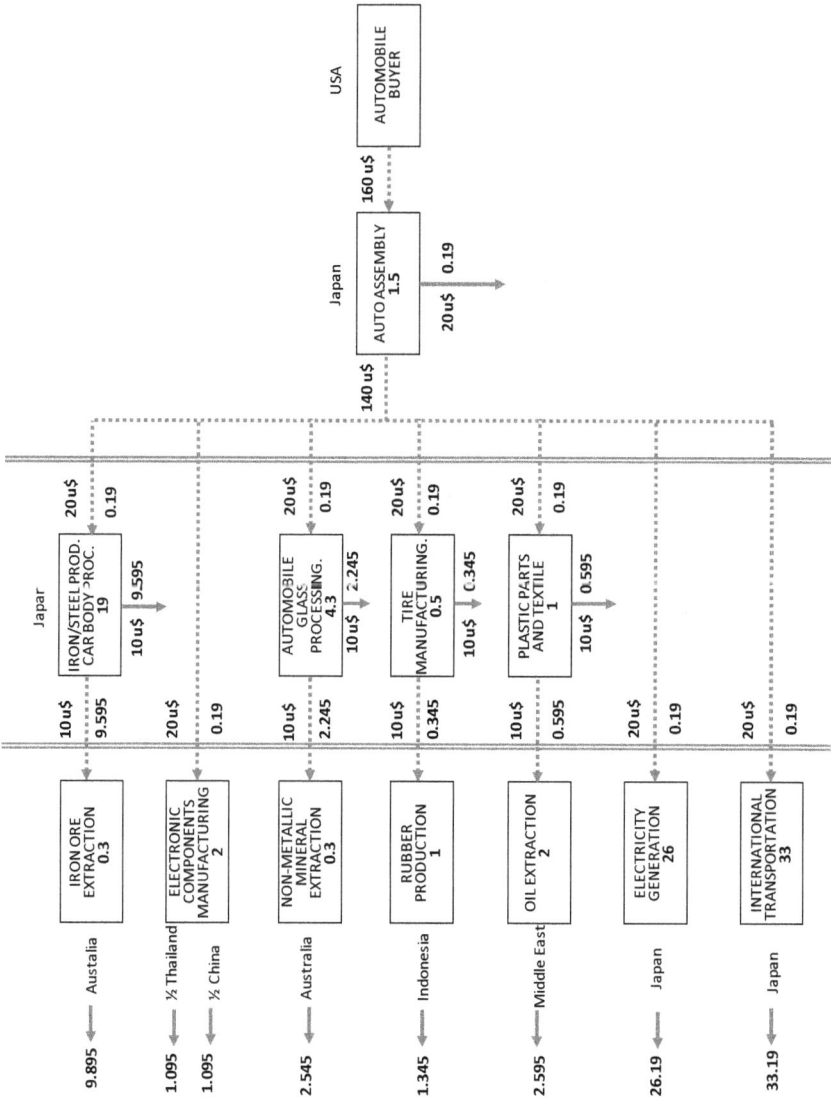

By shared producer and consumer responsibility, an agent acts as both a producer and a consumer at the same time, except for those being at the very beginning of the supply chain which performs only as a producer and those being at the very end of the supply chain which acts only as a consumer. For the example in Box 1, the iron/steel and car body producing sector is a producer which provides car body to the automobile assembly industry and at the same time a consumer which purchases iron ore from the iron ore extraction sector. In this shared producer and consumer responsibility approach, consumer responsibility is different from the consumer responsibility principle as discussed above; since here it is not only the final consumer but also intermediate consumers who assume part of the responsibility. As shown in Figure 4, if a half-half allocation is used, the iron/steel and car body producing sector (in Japan) is allocated 50% of the emissions from its direct upstream supplier, which are added to its own direct emissions and then allocates half of them down to its direct customer, the automobile assembly sector. The emissions accounted for by the iron/steel and car body producing sector is then (0.3/2+19)/2=9.575 units. At the same time, the iron ore extraction (in Australia) is responsible for 0.3/2=0.15 units and the automobile assembly sector (in Japan) is responsible for 9.575 units of emissions. In the same way, two electronic components manufacturers are responsible for 0.5 units by each (Thailand and China) and pass on 1 unit to the automobile assembly industry. Non-metallic mining in Australia is responsible for 0.15 units and passes on 0.15 units to its direct downstream customer, the glass processing sector. The glass processing sector (in Japan) is then responsible for (0.15+4.3)/2=2.225 units and passes another 2.225 units to the automobile assembly sector. Rubber production is responsible for 0.5 units (in Indonesia) and passes on 0.5 units to the tire manufacturing sector, which is finally responsible for (0.5+0.5)/2=0.5 units (Japan) and passes on 0.5 units to the automobile assembly sector. Oil production is responsible for 1 unit (in the Middle East) and passes on 1 unit to the plastic and textile manufacturing sector. The plastic and textile manufacturing sector is then responsible for (1+1)/2=1 unit (in Japan) and passes on 1 unit to the automobile assemble sector. Electricity generation is responsible for 13 units (in Japan) and leaves another 13 units to the account of the automobile assembly sector. International transportation is responsible for 16.5 units (in Japan) and passes on another 16.5 units to the car assembly sector. The car assembly sector (in Japan) is then responsible for [(9.575+1+2.225+0.5+1+13+16.5)+1.5]/2=22.65 units of emissions and passes on another 22.65 units to the consumer (in the USA). For their national inventory, Australia is responsible for 0.15+0.15=0.3 units, Thailand, China and Indonesia are responsible for 0.5 units each, Middle East countries are responsible for 1 unit, Japan is responsible for 9.575+2.225+0.5+1+22.65+13+16.5=65.45 units and the USA is responsible for 22.65 units.

Figure 4: Shared producer and consumer responsibility based on a 50%-50% allocation

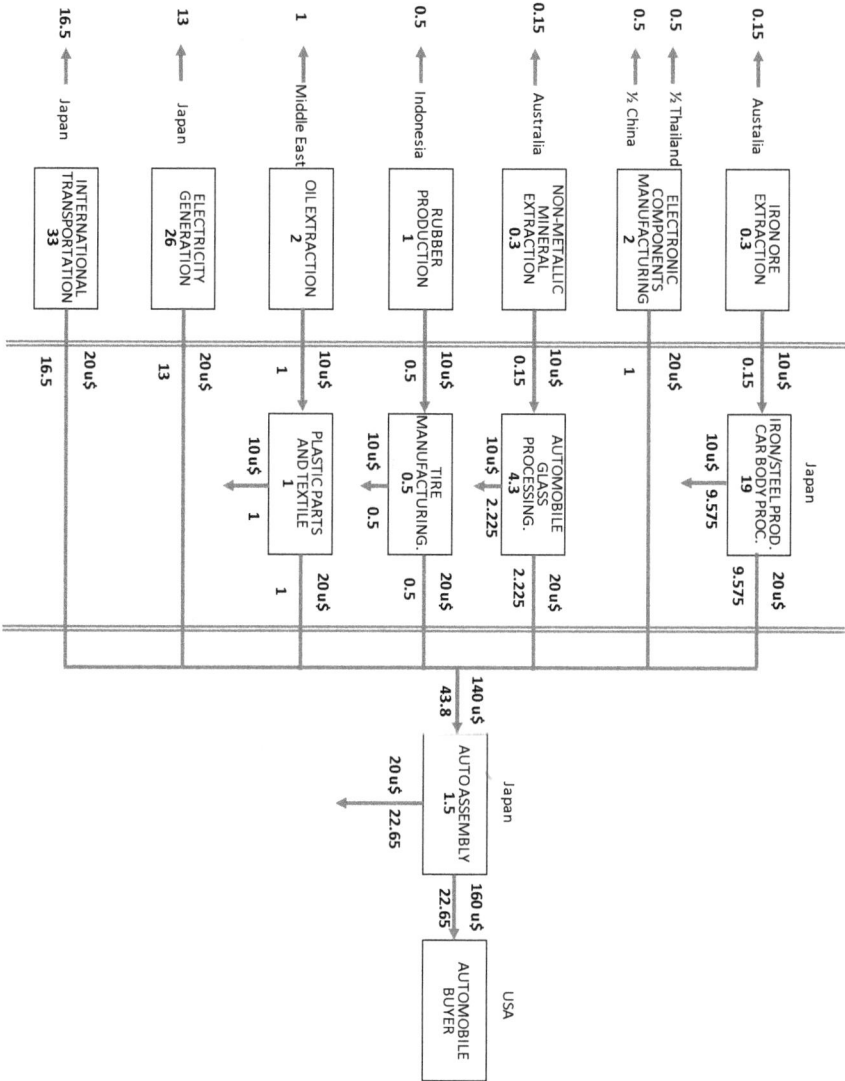

It should be noted that the 50%-50% allocation (or similar allocations such as 40%-60%, etc.) can cause inconsistency in accounting for the responsibilities assumed by the final consumers if the supply chains are arbitrarily broken up into more disaggregated stages (see Lenzen, et al., 2007 for more details), e.g. one of the sectors in the supply chain being split into two or more sectors. A solution suggested by Lenzen and his colleagues to solve this problem is to use a ratio that is independent of sector classification. Value added is such a candidate in that no matter whether a supply chain is represented in a shorter or longer fashion, total value added is always the same at the end of the chain.

In a shared consumer and income responsibility approach, an agent assumes the responsibility as either a final consumer or a provider of primary factors, or as both. Shared consumer and income responsibility is calculated as 1/2 consumer responsibility plus 1/2 income responsibility for each agent. In Figure 2, we see that a buyer of the automobile in the USA, who does not provide any primary factors for this particular supply chain, holds only consumer responsibility but no income responsibility. In Figure 3, we see that the providers of primary factors for all the producing sectors in different countries along the supply chain assume income responsibilities but no consumer responsibilities since they do not consume the particular cars. So, in this supply chain, the shared consumer and income responsibility of each producing sector is 1/2 of its income responsibilities and that of the automobile buyer in the USA is 1/2 of its consumer responsibility. In Figure 5, we show the shared responsibility for all agents involved in the supply chain depicted in Box 1.

On the one hand, the shared consumer and income responsibility approach tries to address the whole supply chain, from the very beginning of the supply chain to the final consumer, by allocating the responsibilities among final consumers and all providers of the primary factors for the supply chain. Normally, a supply chain is a cascade of events where agents contribute in order to deliver goods to final demand. Therefore, in this type of approach, consumer responsibility is concentrated in the final consumer (Figure 2), whereas income responsibility is spread along all the providers of primary factors that contributed to the supply chain (Figure 3).

On the other hand, shared producer and consumer responsibility approach also tries to address the whole supply chain by allocating responsibilities among all agents, each of which performs as both a producer and a consumer. Typically, agents in the beginning of the supply chain will have only producer responsibilities; intermediate sectors will hold both responsibilities and final consumers will hold only consumer responsibilities (Figure 4).

The income-based responsibility principle and the consumer responsibility principle adopt a similar fashion of allocation (embodied emissions), but in opposite directions: downstream in the former case and upstream in the latter case. In the income-based responsibility scheme, downstream responsibilities will be shared among the providers of the primary factors based on their income. In particular, the providers of the primary factors for the very beginning of the supply chain will hold income responsibility for all of its own direct emissions and all downstream emissions enabled by the sector. The providers of the primary factors for an intermediate sector in the supply chain will share its own direct emissions and all downstream emissions enabled by the sector with its upstream suppliers (both direct and indirect). In the consumer responsibility scheme, all upstream responsibilities are allocated to the final consumers.

Figure 5: Shared income and consumer responsibility. The monetary fluxes are not shown but they are equal to those presented in previous figures.

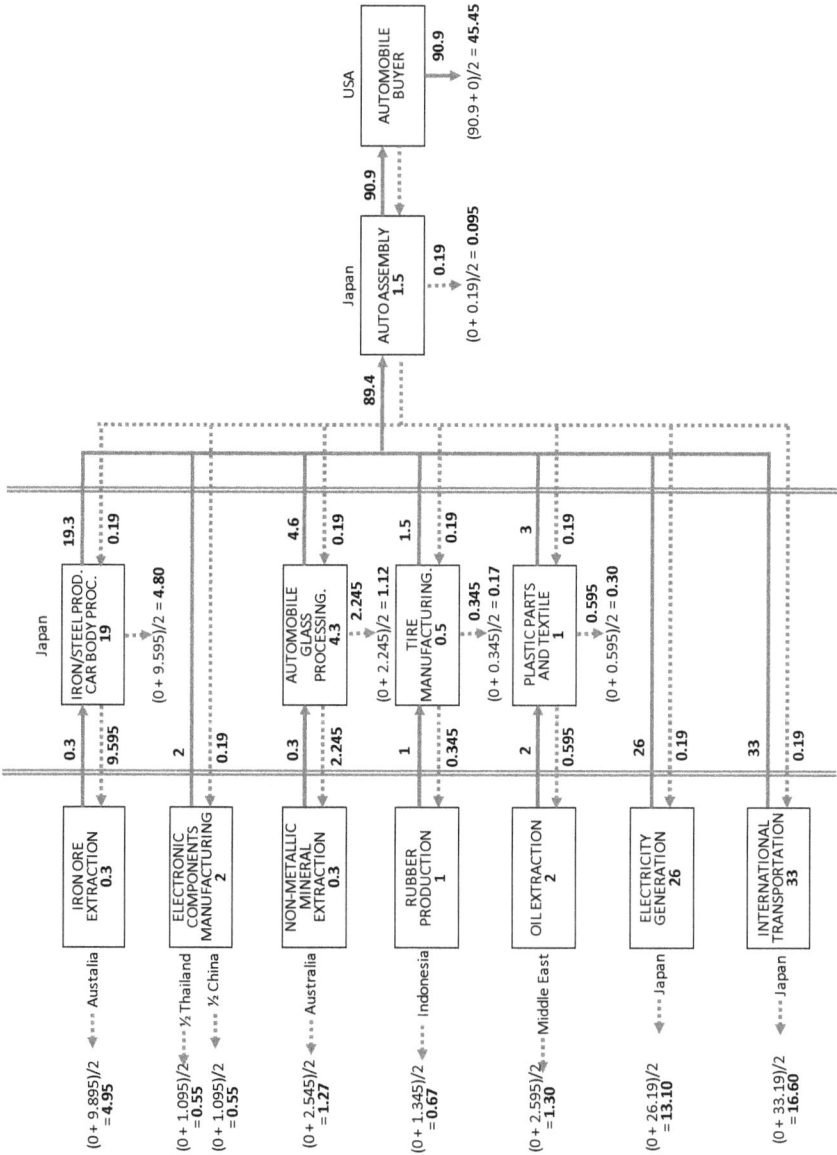

The choice between shared producer and consumer responsibility or shared income and consumer responsibility will depend on the purpose of the study. The main purpose of the former is to share responsibility between firms (intermediate producers) and final consumers, whereas that of the latter is to determine the responsibility of economic agents in their roles as either final consumers or as providers of primary inputs, or as both[6].

Results

Using the AIO 2000, we calculate the national inventories for ten economies based on different responsibility indicators. The results are shown in Table 1.

Table 1: National inventories based on different responsibility indicators

	PRODUCER – TRADE WITH ROW			UPSTREAM EMBODIED EMISSIONS		DOWNSTREAM ENABLED EMISSIONS	
	PRODUCER	PRODUCER-EXPORTS	PRODUCER-IMPORTS	CONSUMER	EXPORTS	INCOME	IMPORTS
INDONESIA	220	182	197	137	38	210	23
MALAYSIA	102	73	83	54	29	85	19
PHILIPPINES	52	43	43	39	9	42	9
SINGAPORE	55	38	30	43	17	39	25
THAILAND	133	104	118	98	29	119	16
CHINA	2865	2523	2632	2261	342	2495	233
TAIWAN	161	118	136	108	43	151	25
KOREA	347	277	279	278	70	296	68
JAPAN	869	815	818	944	54	889	51
USA	4598	4269	4176	4481	329	4186	422
SUB-TOTAL				8442	962	8512	892
	9404	8442	8512				
TOTAL				9404		9404	

NOTE: In this analysis we opted to leave out international trade with the rest of the world (ROW), therefore in the computations, emissions embodied in imports and emissions generated in ROW that are enabled by exports of the ten economies to ROW are not included. For producer responsibility we have computed 3 quantities: producer, producer – exports and producer – imports. The first respects the direct emissions taking place within each country; producer – exports respects the upstream emissions embodied in exports to ROW subtracted from direct emissions; and producer – imports respects the downstream emissions enabled by imports from ROW subtracted from direct emissions. These adjustments must be done in order to compare producer with consumer and income responsibility that do not include trade with ROW. Consumer responsibility concerns emissions embodied in each country's final demand, and income responsibility concerns emissions enabled by each country's primary suppliers. The areas shaded are the values analysed in this work. For details, please refer to Rodrigues *et al.* (2006) and Zhou, *et al.* (2010).

Producer vs. consumer responsibility

Regarding the 10 economies described by the AIO 2000, we see that irrespective of the type of responsibility analysed the USA always accounts for more than

---

[6] Lenzen, M. (2008). Consumer and producer environmental responsibility: A reply. *Ecological Economics*, 66, 547-550.

50% of total responsibility, followed by China with approximately 30%, and Japan with approximately 10% of total responsibility.

While producer responsibility provides information about the emissions that occur due to the production processes of a country, consumer responsibility provides information about the emissions that are generated directly and indirectly due to the final consumption of a country. For a country the difference between these two types of responsibility occurs because of international trade[7]. If a country holds a consumer responsibility higher than its producer responsibility (producer-exports in Table 1), this indicates that the emissions embodied in its imports, in order to satisfy its demand, are higher than the emissions embodied in its exports. We see that the USA, Japan, Singapore and Korea have a higher consumer responsibility than producer responsibility, whilst for the USA and Japan the difference is considerable 212 and 129 $MtCO_2$, for Singapore and Korea the difference is very small 5 and 1 $MtCO_2$ (Table 1). Nevertheless, we see that higher income countries are also those with higher differences between consumer and producer responsibilities. The USA and Japan's final demand generated more emissions than their direct emissions. In terms of climate policy, this indicates that consumption in these countries generates more emissions than the emissions that are generated within their borders. Therefore if consumer responsibility were to be applied in climate policy, these countries would have an increased responsibility. China and Indonesia are the countries where the differences between producer and consumer responsibilities are higher. This indicates that part of the direct emissions generated through production processes in these countries occur to fulfil the demand in other countries. If consumer responsibility were to be applied in climate policy, these countries would have a decreased responsibility.

Producer vs. income responsibility

As seen before, producer responsibility informs about the emissions that occur due to the production processes occurring within that country; income responsibility informs about the emissions that were directly and indirectly enabled by the providers of primary factors in exchange for an economic benefit. For a country, the differences in these quantities occur due to international trade. If a country holds an income responsibility higher than its producer responsibility (producer-imports in Table 1), this indicates that the overseas emissions enabled by its exports are higher than the emissions enabled domestically by its imports. Or in other words, that in order to generate its value added, a country enables emissions elsewhere.

Here we see (Table 2) that China and the Philippines are the only countries that have an income responsibility smaller than their producer responsibility. So, part of the emissions occurring within their borders is to provide economic benefit to the primary suppliers of other countries. Singapore and Taiwan are the

---

[7] Kanemoto, K., Lenzen, M., Peters, G., Moran, D., & Geschke, A. (2012). Frameworks for comparing emissions associated with production, consumption and international trade. *Environmental Science and Technology* 46, 172-179.

countries with a highest percentage increase in their income responsibility when compared to the producer responsibility. This means, that the income of these countries depends on emissions enabled in foreign territories.

Table 2: Shared responsibility indicators

	SHARED PRODUCER AND CONSUMER	SHARED INCOME AND CONSUMER
INDONESIA	173	173
MALAYSIA	64	70
PHILIPPINES	42	41
SINGAPORE	40	41
THAILAND	103	108
CHINA	2460	2378
TAIWAN	112	129
KOREA	275	287
JAPAN	850	916
USA	4324	4334
SUB-TOTAL	**8442**	**8477**

Shared responsibility indicators

In the shared producer and consumer responsibility scenario, the direct emissions that each country generates within its borders and the emissions embodied in each country's consumption (both final and intermediate) are analysed. In the shared income and consumer responsibility, the emissions that a country enables through the provision of primary factors and the emissions embodied in its final consumption are analysed. Interestingly, we see that both shared approaches yield similar results. If we compared these results with the metric used by the Kyoto Protocol (Table 1 – Column producer-exports), in both approaches Indonesia, Malaysia, Taiwan and China have a decreased responsibility, whereas other countries have an increased responsibility. In some cases, the differences are very small, namely for Philippines, Thailand and Taiwan. Nevertheless, in the shared consumer and income approach there are bigger differences.

Conclusions

With the rapid development of emerging economies such as China and India, to tackle climate change without the participation of large developing emitters will remain very difficult and costly. Though developing countries have committed to take voluntary actions, namely nationally appropriate mitigation actions (NAMA), there is high pressure that they should commit more to quantified reductions. Many large emitters from the developing world attribute their growth to exports. They argue that emissions remain in their national inventories while the importing countries, especially developed countries, enjoy the benefits from consumption and reduce their national inventories without necessarily making substantial efforts. As China's top climate change negotiator, Li Gao, said, China should not pay for cutting emissions caused by the high consumption of other countries. He therefore calls for a fairer agreement to address emissions embodied in international trade[8] (BBC, 2009).

To address the problems related to producer responsibility adopted in the national inventory, we provided several alternative indicators on the basis of which we compared the national emissions by using the AIO 2000 for the case of Asian countries. Results indicate that a change from production-based accounting to other accounting principles will influence national emissions significantly. The consequences are profound for individual countries.

Though accounting for emissions related to international trade is an important issue, it is yet to be put onto the agenda of the UNFCCC for serious consideration. Several reasons can explain this. First, a change in the accounting method from production-based inventory to consumption or income-based inventory as shown in our case study, will influence the amount of emissions each country is responsible for and as a result will substantially influence quantified mitigation targets. There are both winners and losers. To reach an agreement among parties will be difficult. Second, by the consumption or income responsibility, a big political challenge is that the boundary of environmental responsibility is not consistent with the jurisdiction of a country that commits to control and limit the emissions. Third, a change from full producer responsibility to full consumer or income responsibility cannot solve the equity problem. In particular, the consumer usually has limited choices over technologies used in the production, which cause the emissions. In this sense, a shared environmental responsibility is fairer but more complicated for operation.

Regardless of the difficulties, consumption-based accounting can help address the environmental pressures caused by overconsumption and lifestyles. Through environmentally informed purchase, a cascade of demand in good environmental behaviour from the end of the supply chain to the very upstream mining activities can be formed. In addition, by the inclusion of the emissions embodied in imports, it can extend the coverage of emissions stipulated by the Kyoto Protocol and help address the carbon leakage issue. In many senses, consumption-based accounting should be taken into account by the UNFCCC as complementary information in defining national emissions responsibilities and mitigation targets.

Further readings

Gallego, B., & Lenzen, M. (2005). A consistent input-output formulation of shared producer and consumer responsibility. *Economic Systems Research*, 17 (4), 365-391.

---

[8] BBC, 2009. *China seeks exports carbon relief*. Retrieved April 2009 from http://news.bbc.co.uk /2/hi/7947438.stm.

Lenzen, M., & Murray, J. (2010). Conceptualising environmental responsibility. *Ecological Economics*, 70, 261-270.

Lenzen, M., Murray, J., Sack, F., & Wiedmann, T. (2007). Shared producer and consumer responsibility – Theory and practice. *Ecological Economics*, 61, 27-42.

Marques, A.; Rodrigues, J.; Lenzen, M.; Domingos, T.. (2012) Income-based environmental responsibility. *Ecological Economics,* 84, 57-65.

Rodrigues, J., Domingos, T., Giljum, S., & Schneider, F. (2006). Designing an indicator of environmental responsibility. *Ecological Economics*, 59, 256-266.

Rodrigues, J., & Domingos, T. (2008). Consumer and producer environmental responsibility: Comparing two approaches. *Ecological Economics*, 66, 533-546.

Zhou, X., Liu, X.B., & Kojima, S. (2010). *Carbon Emissions Embodied in International Trade: An Assessment from the Asian Perspective.* Hayama: Institute for Global Environmental Strategies (IGES). Retrieved April 2012 from http://enviroscope.iges.or.jp/modules/envirolib/upload/2719/attach/embodied_emissions_whole_sf_publication.pdf.pdf

Acknowledgments

Xin Zhou would like to thank the Japan Society for the Promotion of Science for its Grants-in-Aid for Scientific Research ((B) No. 21310033). Alexandra Marques would like to thank the financial support of FCT and the MIT Portugal program through scholarship SFRH/BD/42491/2007, and the valuable comments and suggestions of Tiago Domingos and João Rodrigues. We also express our sincere thanks to Joy Murray and Manfred Lenzen for their valuable comments and careful editing work, which help make the expressions and explanations in the chapter more precise and clearer.

Chapter 17: Sustainability Assessment from a Global Perspective with the EXIOPOL Database

Richard Wood and Kjartan Steen-Olsen

Background and context

The collective goal we have to improve livelihoods – and consequently create a better world – has historically often been interpreted as a need for increased economic development across the globe. In more recent times, however, there has been an increasing awareness that economic development may not be wholly synonymous with that fundamental goal of 'a better world'. Some economists have argued that economic development will *eventually* provide increased wealth for all, as well as providing the ability to deliver healthy social structures and allow protection of the environment. However, there is little evidence that economic development eventually decelerates all negative impacts (see literature on the environmental Kuznets hypothesis). Consequently we are now seeing attention being paid to concerns beyond economic growth with development goals increasingly taking into account social good and environmental protection. For example, governments are preserving land for non-industrial uses that previously would have been destroyed in the name of economic development. Faced with problems that will result from predicted anthropogenic-induced climate change, governments are now trying to reduce greenhouse gas emissions, even at the expense of economic growth.

However, the problems are intertwined. In advancing economic development we have polluted our air. In trying to address air pollution, we have started clearing our forests for biofuels, bringing a whole new set of problems. It seems that when trying to solve complex problems by breaking them down into single issues we face the risk of *solving one problem, but creating many more*. Not only might we create more problems, but when we take a local solution to the problem, we may also end up shifting the problem elsewhere.

MRIO has played a vital role in investigating trans-border impacts; no longer can we get away with putting our own house in order without worrying about our neighbours. The analyses shown in this book trace impacts along the supply chain and across country borders. MRIO is a very useful tool for examining impacts driven by the consumption of goods and services, but we still need to be mindful of the breadth of impacts we are trying to address.

As mentioned, in addressing one type of impact, we risk creating other types of completely different impacts. In practice there will always be trade-offs involved when politicians form policy, be it the trade-off between reducing greenhouse gas emissions and protecting employment or between reducing greenhouse gas emissions and increasing land clearing. To account for this trade-off, our analysis needs to go beyond simple single indicator assessments. In this chapter we focus on this challenge by first giving a global perspective of the issues of economic development, employment and greenhouse gas emissions. We follow on from the global economy-wide perspective with an analysis of consumption by product in the European Union (EU) according to a range of indicators to investigate the trade-offs that might be relevant to consider.

Process

The construction of the EXIOPOL database involved many people and institutions, and is outlined in Chapter 6 of this book. The objective of the database was to allow analysis of environmentally important issues from a global perspective, with emphasis on the European Union. As such the main concerns were 1) to provide consistency and tractability of both production structures and consumption activities in major world regions, and 2) to provide a broad coverage of environmental flows. These flows included emissions to air, water, and land, as well as inputs from nature, such as water extraction, land use, and material extraction. The EXIOPOL database was created under the United Nations' System of National Accounts with a consistent representation of impacts across the environmental flows, as well as across the social indicator of employment and a range of economic indicators. This is one of the strengths of MRIO analysis – we have a consistent framework to include economic, social and environmental impacts.

As with the other analyses described in this book, we trace a number of impacts from primary cause, through secondary production and possibly trade, to final consumption of goods and services by the populace.

In this chapter, we show the expedient nature of this approach using the EXIOPOL database. We focus on a sustainability assessment including results and analysis across the three pillars of environmental, social and economic sustainability. In order to make policy relevant for sustainable production and consumption, it is insightful to know the trade-offs between the different pillars of sustainability for the basket of goods and services consumed. Furthermore, when focussing on issues like reducing greenhouse gas emissions, it is important that we are not causing other environmental problems. We thus include a range of environmental metrics, from macro-resources such as land and water use, to impact weighted measures of toxic flows. Often environmental metrics focus on large-scale environmentally relevant flows, but the impact that an emission may have on the environment is often most pronounced following small releases of highly toxic substances (such as radioactive material). Hence, with the EXIOPOL data, we include life-cycle mid-point indicators for freshwater eco-toxicity, freshwater eutrophication, and terrestrial acidification. These indicators are frequently included in the impact assessment stage of life-cycle analyses, and we provide suggested further reading at the end of this chapter.

We focus on the consumption of products within the EU. Whilst analyses of individual countries within the MRIO framework are definitely interesting in themselves, we choose to focus on product consumption at the EU level so that product policy can be designed at that level. The EU represents one of the most significant groups of developed countries in the world, with large implications for other regions because of trade-related production linkages.

Standard methods of MRIO analysis are employed (see Section 1 and 2), and we will not repeat the derivations of those methods here. We take the footprint approach to show impacts by indicator by country rather than the impacts embodied in trade approach. This ensures that all life-cycle emissions are correctly allocated to final products, even though the intermediate stages of production associated with trade will necessarily be double counted if trade balances are explicitly investigated.

Fig. 1 compares European countries and their main trading partners in terms of three quite different indicators, each relevant to one of the sustainability dimensions. Along with the footprint accounts, matching territorial accounts are also shown for each indicator. Compared to the footprint accounts, which take the consumption perspective, the territorial accounts take the production perspective, counting the total amounts occurring within the borders of each country. Each value is shown relative to the global average on a logarithmic scale, and countries are sorted by affluence, represented by GDP per capita.

The countries vary widely in terms of value added per capita, with Luxembourg about two orders of magnitude higher than India. The variation is smaller for the employment and emission indicators, although there is a tendency for more affluent countries to have higher footprints. Some countries, like Russia and Greece, emit a lot of greenhouse gases even though their consumption does not lead to correspondingly high emissions. This indicates that their exports carry higher embodied emissions than their imports. This can be a result of more emission intensive products being exported, overall higher export than import volumes, or both. Denmark, Canada, Australia, Greece, Czech Republic, Estonia, Slovakia, South Africa, and the bottom six countries (including Russia, China, and India) all have an emissions trade surplus – they export more embodied greenhouse gas emissions than they import. In contrast, most wealthy countries, such as Japan, Switzerland, Sweden, and the UK, all have embodied greenhouse gas emission trade deficits, which means that they import products with more embodied emissions than they export.

Results

Figure 1: Territorial impacts (dotted line) and footprints (solid line) per-capita relative to world average.

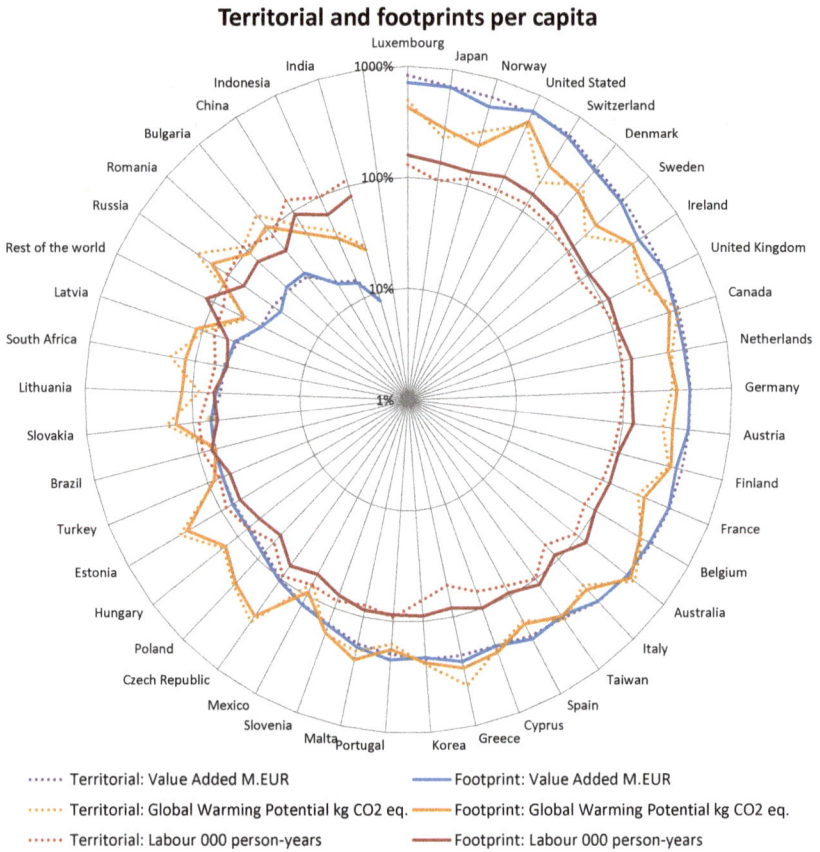

Territorial and footprints per capita

The presentation of the territorial value added versus footprint value added gives a direct comparison of trade balances. Territorial value added sums to gross domestic product, which is equivalent to the sum of the economic value of domestic consumption, minus value of exported goods plus value of imported goods. The footprint of value added includes the sum of upstream impacts for domestic consumption plus impacts for imported goods, minus impacts for exported goods. What is interesting when analysing value added, is that because of the duality of the input-output representation, the value added multipliers are necessarily equal to the price of a product (that is, the price is equivalent to the summation of all value added over the supply chain). In economic units, the price of a product is simply the unit price (i.e. 1). Hence our value added footprint reduces to domestic consumption plus imports minus exports, which is an alternative measure of gross national income. We see that countries like

Luxembourg, Norway and Ireland have large positive trade balances, whilst countries like the United States and Greece have negative trade balances.

The 25 or so most affluent countries in Fig. 1 all have higher labour footprints per capita than the territorial equivalent. This means that the total consumption by the average person in these countries requires more man-hours of labour than the amount of hours worked by this person. Richer countries have higher salary levels and so tend to specialize in technology intensive production, while the labour intensive industry is more often located in countries where labour is cheaper.

Following the principle of consumer responsibility, every production activity and every impact from these activities is driven by the final demand of goods and services by households and governments. The strength of input-output analysis is its ability to allocate observed impacts to specific product components of the household and government final demand. The red bubbles in Fig. 2 quantitatively show how much the demand for each product included in the EXIOPOL system contributes to the global total for eight types of impacts. Here we expand our list of indicators to take advantage of the coverage of environmental impacts in the EXIOPOL database in order to try and prioritise products that do not cause problem shifting – e.g. products that can help reduce greenhouse gas emissions, without causing eutrophication or acidification instead.

In the upper part we see that purchases of products from the *Crops not elsewhere classified (nec)* group contribute most to global land transformation. Purchases of metal ores on the other hand contribute almost nothing to any of the impact types. This is because the total contribution of a product is determined by its multiplier, or the total impact caused per unit of demand, as well as the total amount demanded by final consumers (such as households). While metal ores are essential inputs to the global economy, they are very rarely purchased directly by households.

Following this reasoning, a disaggregation of the total impact is of interest. The blue bubbles in Fig. 2 show the intensity components of the total impact of each product group. These measures are generally referred to as multipliers, and they represent the total life-cycle impacts caused by purchasing $1 of a specific product. This part of Fig. 2 reveals that one dollar spent on products from the forestry and logging sector causes much more land transformation than a dollar spent on product from the other crops sector, even though purchases of the latter contribute more to land transformation globally. The disaggregation also shows another interesting result: purchases of hotel and restaurant services (see page three of Fig. 2) contribute substantially in terms of many of our global environmental concerns, nevertheless choosing to spend money on such services can hardly be said to be a bad choice for the environmentally concerned consumer as the impact per $ spent is small.

In prioritising products for the implementation of environmental policies the results show that energy use is important, with electricity and fuels showing the highest embodied emissions (here even excluding emissions from direct household fuel use). Regarding water and land use, agriculture and food products clearly have higher impacts.

	% of total Impact: Employment: 000 person-years	% of total Impact: Freshwater ecotoxicity: kg 1,4-DCB-Eq	% of total Impact: Freshwater eutrophication: kg P-Eq	% of total Impact: Global warming (GWP100 (IPCC, 2007)): kg CO2 eq.	% of total Impact: Natural land transformation : m2	% of total Impact: Terrestrial acidification: kg SO2-Eq	% of total Impact: Value Added: M.EUR
Paddy rice	·	·	·	·	·	·	·
Wheat	•	•	●	●	●	•	·
Cereal grains nec	•	•	●	●	●	•	·
Vegetables, fruit, nuts	●	●	●	●	●	●	●
Oil seeds	·	•	●	•	●	·	·
Sugar cane, sugar beet	·	·	·	·	·	·	·
Plant-based fibers	·	·	·	·	·	·	·
Crops nec	•	●	●	●	⬤	●	●
Cattle	·	·	●	·	•	●	·
Pigs	·	·	●	·	·	·	·
Poultry	•	•	●	●	●	●	●
Meat animals nec	·	·	·	·	·	·	·
Animal products nec	·	·	●	·	·	•	·
Raw milk	●	•	●	●	●	●	·
Wool, silk-worm cocoons	·	·	·	·	·	·	·
Products of forestry, logging ..	•	·	·	·	●	·	·
Fish and other fishing produ..	•	·	·	·	•	·	·
Coal and lignite; peat (10)	·	·	·	·	·	·	·
Crude petroleum and service..	·	·	·	·	·	·	·
Natural gas and services rel..	·	·	·	·	·	·	·
Other petroleum and gaseou..	·	·	·	·	·	·	·
Uranium and thorium ores (1..	·	·	·	·	·	·	·
Iron ores	·	·	·	·	·	·	·
Copper ores and concentrat..	·	·	·	·	·	·	·
Nickel ores and concentrates	·	·	·	·	·	·	·
Aluminium ores and concent..	·	·	·	·	·	·	·
Precious metal ores and con..	·	·	·	·	·	·	·
Lead, zinc and tin ores and c..	·	·	·	·	·	·	·
Other non-ferrous metal ores..	·	·	·	·	·	·	·
Stone	·	·	·	·	·	·	·
Sand and clay	·	·	·	·	·	·	·
Chemical and fertilizer miner..	·	·	·	·	·	·	·
Products of meat cattle	•	●	⬤	●	●	⬤	●
Products of meat pigs	•	●	⬤	●	●	⬤	●
Products of meat poultry	●	●	⬤	●	●	●	●
Meat products nec	•	●	●	●	●	●	●
products of Vegetable oils a..	•	•	●	●	●	•	·
Dairy products	●	●	⬤	●	●	⬤	●
Processed rice	·	·	·	·	·	·	·
Sugar	•	•	•	·	•	•	●
Food products nec	●	●	●	●	⬤	●	●
Beverages	·	•	•	●	•	·	·
Fish products	●	●	●	●	•	●	●

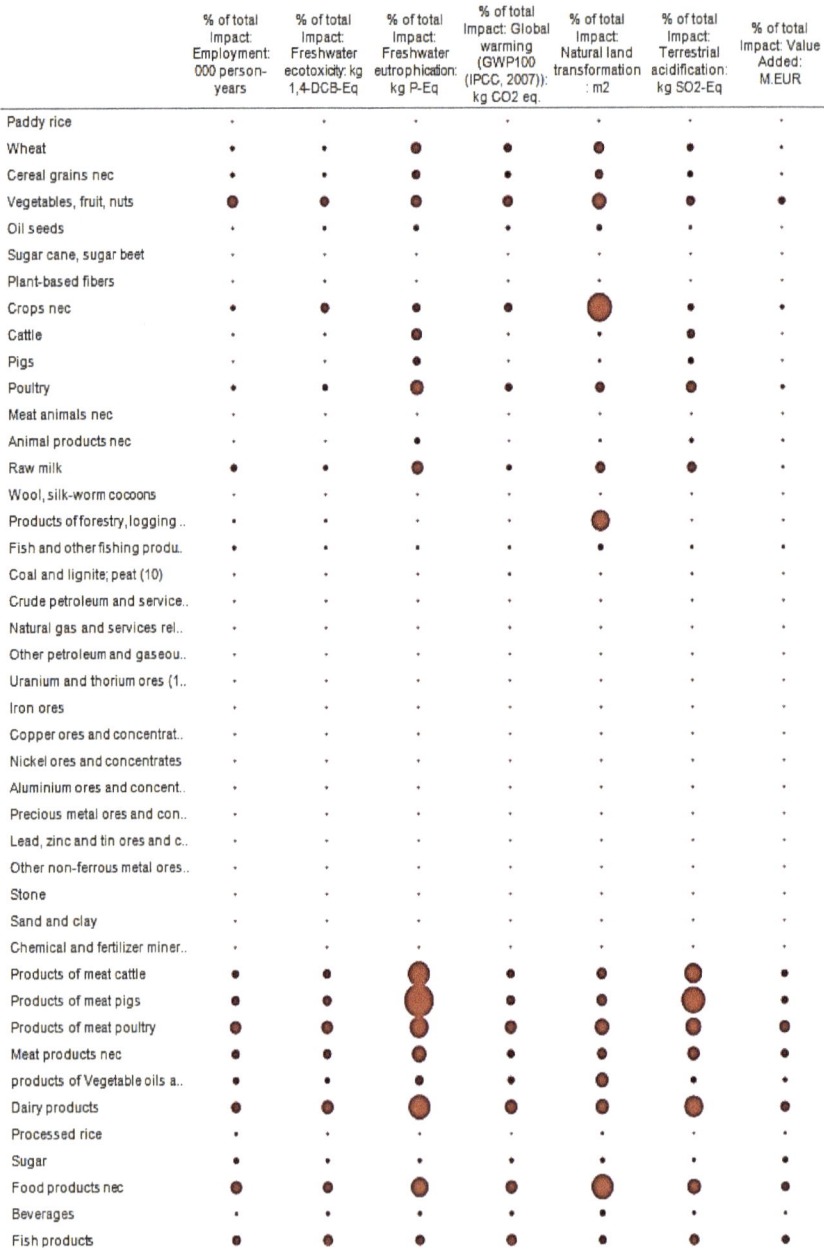

Table 1: Left: Percentage of total impact embodied in EU final consumption by impact category and product group; Right: Multipliers compared to average multiplier by impact category and product group.

% of total Impact: Water depletion: m3	Multipliers: Employment: 000 person-years	Multipliers: Freshwater ecotoxicity: kg 1,4-DCB-Eq	Multipliers: Freshwater eutrophication : kg P-Eq	Multipliers: Global warming (GWP100 (IPCC, 2007)): kg CO2 eq.	Multipliers: Natural land transformation : m2	Multipliers: Terrestrial acidification: kg SO2-Eq	Multipliers: Value Added: M.EUR	Multipliers: Water depletion: m3

	% of total Impact: Employment: 000 person-years	% of total Impact: Freshwater ecotoxicity: kg 1,4-DCB-Eq	% of total Impact: Freshwater eutrophication: kg P-Eq	% of total Impact: Global warming (GWP100 (IPCC, 2007)): kg CO2 eq.	% of total Impact: Natural land transformation: m2	% of total Impact: Terrestrial acidification: kg SO2-Eq	% of total Impact: Value Added: M.EUR
Tobacco products (16)	·	·	·	·	●	·	·
Textiles (17)	●	●	•	•	●	●	•
Wearing apparel; furs (18)	⬤	●	•	●	●	●	●
Leather and leather products..	⬤	•	•	·	•	•	•
Wood and products of wood ..	·	·	·	·	●	·	·
Pulp, paper and paper produ..	·	•	•	•	●	•	•
Printed matter and recorded ..	•	•	•	•	●	•	●
Coke oven products	·	·	·	·	·	·	·
Motor spirit (gasoline)	⬤	•	·	●	·	·	·
Kerosene, including kerosen..	·	·	·	·	·	·	·
Gas oils	•	·	·	•	·	·	·
Fuel oils n#e#c#	·	•	·	·	·	·	·
Petroleum gases and other g..	·	·	·	·	·	·	·
Other petroleum products	·	·	·	·	·	·	·
Nuclear fuel	·	·	·	·	·	·	·
Chemicals, chemical product..	●	⬤	•	●	•	●	●
Rubber and plastic products ..	•	•	·	•	·	•	•
Glass and glass products	·	•	·	•	·	•	·
Ceramic goods	·	•	·	•	·	•	·
Bricks, tiles and construction..	·	·	·	·	·	·	·
Cement, lime and plaster	·	·	·	·	·	·	·
Other non-metallic mineral p..	·	•	·	·	·	·	·
Basic iron and steel and of f..	·	•	·	•	·	·	·
Precious metals	·	·	·	·	·	·	·
Aluminium and aluminium pr..	·	·	·	·	·	·	·
Lead, zinc and tin and produ..	·	·	·	·	·	·	·
Copper products	·	·	·	·	·	·	·
Other non-ferrous metal pro..	·	·	·	·	·	·	·
Foundry work services	·	·	·	·	·	·	·
Fabricated metal products, e..	•	•	·	•	·	•	•
Machinery and equipment n..	•	•	·	•	·	•	●
Office machinery and compu..	•	·	·	•	·	·	•
Electrical machinery and app..	•	•	·	•	·	·	•
Radio, television and commu..	●	•	·	•	·	•	●
Medical, precision and optic..	•	·	·	•	·	·	•
Motor vehicles, trailers and s..	⬤	⬤	•	●	•	●	⬤
Other transport equipment (3..	•	•	·	·	·	·	•
Furniture; other manufacture..	⬤	⬤	•	●	●	●	⬤
Metal secondary raw materia..	·	·	·	·	·	·	·
Non-metal secondary raw m..	·	·	·	·	·	·	·
Electricity by coal	·	⬤	⬤	⬤	·	⬤	·
Electricity by gas	·	•	·	⬤	·	·	•
Electricity by nuclear	·	·	·	·	·	·	·

% of total Impact: Water depletion: m3	Multipliers: Employment: 000 person-years	Multipliers: Freshwater ecotoxicity: kg 1,4-DCB-Eq	Multipliers: Freshwater eutrophication : kg P-Eq	Multipliers: Global warming (GWP100 (IPCC, 2007)): kg CO2 eq.	Multipliers: Natural land transformation : m2	Multipliers: Terrestrial acidification: kg SO2-Eq	Multipliers: Value Added: M.EUR	Multipliers: Water depletion: m3

	% of total Impact: Employment: 000 person-years	% of total Impact: Freshwater ecotoxicity: kg 1,4-DCB-Eq	% of total Impact: Freshwater eutrophication: kg P-Eq	% of total Impact: Global warming (GWP100 (IPCC, 2007)): kg CO2 eq.	% of total Impact: Natural land transformation: m2	% of total Impact: Terrestrial acidification: kg SO2-Eq	% of total Impact: Value Added: M.EUR
Electricity by hydro	·	·	·	·	·	·	·
Electricity by wind	·	·	·	·	·	·	·
Electricity nec, including bio..	•	●	•	●	·	●	•
Transmission services of ele..	·	·	·	·	·	·	·
Distribution and trade servic..	•	•	·	•	·	·	●
Manufactured gas and distri..	•	·	·	•	·	·	·
Steam and hot water supply ..	•	●	•	●	·	●	•
Collected and purified water,..	•	•	•	•	·	•	•
Construction work (45)	●	●	•	●	•	•	●
Sale, maintenance, repair of ..	●	●	•	•	•	●	●
Retail trade services of moto..	•	·	·	•	·	·	•
Wholesale trade and commi..	●	●	●	●	●	●	●
Retail trade services, except..	⬤	●	●	●	●	●	⬤
Hotel and restaurant service..	⬤	●	⬤	●	⬤	●	⬤
Railway transportation servic..	·	·	·	·	·	·	·
Other land transportation ser..	●	●	•	●	•	●	●
Transportation services via p..	·	·	·	·	·	·	·
Sea and coastal water trans..	•	•	·	●	·	·	•
Inland water transportation s..	·	●	•	•	·	•	·
Air transport services (62)	●	•	•	●	•	•	●
Supporting and auxiliary tran..	●	•	•	●	•	•	●
Post and telecommunication ..	●	●	•	●	•	●	●
Financial intermediation serv..	●	•	·	•	·	•	●
Insurance and pension fundi..	●	•	·	•	·	•	●
Services auxiliary to financial..	•	·	·	·	·	·	•
Real estate services (70)	●	●	●	●	●	●	⬤
Renting services of machine..	•	·	·	·	·	·	•
Computer and related servic..	·	·	·	·	·	·	·
Research and development ..	·	·	·	·	·	·	·
Other business services (74)	•	•	·	•	·	·	●
Public administration and def..	•	•	·	•	·	·	●
Education services (80)	●	•	·	•	·	•	●
Health and social work servi..	●	●	•	●	•	●	●
Collection and treatment ser..	·	·	·	·	·	·	·
Collection of waste	•	•	·	•	·	·	•
Incineration of waste	·	●	·	·	·	·	·
Landfill of waste	·	·	·	·	·	·	·
Sanitation, remediation and ..	·	·	·	·	·	·	·
Membership organisation se..	•	·	·	·	·	·	·
Recreational, cultural and sp..	●	•	•	●	•	•	●
Other services (93)	•	•	·	•	·	•	●
Private households with emp..	●	·	·	·	·	·	●
Extra-territorial organizations..	·	·	·	·	·	·	·

% of total Impact: Water depletion: m3	Multipliers: Employment: 000 person-years	Multipliers: Freshwater ecotoxicity: kg 1,4-DCB-Eq	Multipliers: Freshwater eutrophication : kg P-Eq	Multipliers: Global warming (GWP100 (IPCC, 2007)): kg CO2 eq.	Multipliers: Natural land transformation : m2	Multipliers: Terrestrial acidification: kg SO2-Eq	Multipliers: Value Added: M.EUR	Multipliers: Water depletion: m3

Discussion

Although discussions of sustainability are often spurred by environmental concerns, any approach towards a truly sustainable society must simultaneously address environmental, social and economic matters. The analysis presented here can provide important insights to this end by assessing total global impacts of consumption in terms of a range of indicators. It is clear that we have large disparities in impacts globally across our reported indicators, especially for wealth (GDP) and greenhouse gas emissions. MRIO analysis is not needed for this. However, the MRIO analysis has shown that this inequality is exacerbated especially for greenhouse gas emissions and labour by the international flow of goods and services. Developed countries are benefitting from cheap foreign labour, whilst also causing relatively higher greenhouse gas emissions in foreign countries.

Of great importance to the environmental leaders and decision-makers of the twenty-first century will be the ability to address specific environmental concerns whilst always keeping an eye on the broader picture – minding the indirect effects on the overall system that might result from individual efforts. End-of-pipe solutions to individual environmental problems always suffer risks of shifting rather than solving the problem at hand. Such shifts could be in time, for example if a factory treats a waste stream output simply by diluting it, or in space, if domestic legislation causes polluting processes to be relocated to countries with less strict environmental regulations. Finally the problem can be shifted to other types of environmental issues, as illustrated by the land use concerns now arising from the transition to biofuels introduced to reduce greenhouse gas emissions.

The challenge for researchers will be to accurately model these large and complex systems, and equally importantly, to communicate results to governments and companies. MRIO analysis is a tool that is well suited to do this and the increased coverage of environmental stressors in modern databases allows for detailed analyses of trade-offs between various environmental issues.

Consumers have the ability to modify product choices in order to influence the production patterns of producers. Using MRIO analysis to look at life-cycle impacts allows for a more complete picture to be made. As an example, consider a furniture manufacturer faced with a choice of whether to use textile or leather upholstery for a new range of sofas. Wishing to green their products, the management is leaning towards leather, having heard concerns about the large freshwater requirements from cotton production. If we look back at our results, and compare the products *Textiles* and *Leather and leather products* in Fig. 2, this notion is confirmed, however the results also show that leather might negatively affect freshwater systems another way, through eutrophication. Hence the impacts and trade-offs between individual decisions on the broader system can be immediately captured. This is (hopefully) where we are headed in designing policy for sustainable production and consumption.

Conclusions

In pursuing development goals for the 21st century, we are increasingly seeing the need for policies to be steered away from economic growth as the sole criterion, and towards a holistic view of sustainable development. In pursuing sustainable development, we are increasingly seeing the need for a global approach in order to avoid problem shifting, and such an approach can be captured by MRIO analysis. Consumption in one country is inextricably linked to environmental impacts, wealth creation and labour use in other countries. The large inequalities that exist around the world are being exacerbated by trade, where cheap labour and lenient emissions regulations are being exploited by the wealthiest countries.

In trying to derive policy and drive consumer action on these issues, we need to make consumers aware of the total life-cycle impact of their consumption. Only by doing so can we ensure that informed choices can be made, and that the problems we strive so hard to solve are not simply shifted from one place to another.

Further reading

Goedkoop, M., Heijungs, R., Huijbregts, M., De Schryber, A., Struijs, J., & Van Zelm, R. (2008). A life cycle impact assessment method which comprises harmonised category indicators at the midpoint and the endpoint level; First edition Report I: Characterisation. Retrieved January 6, 2009 from ReCiPe 2008: http://www.lcia-recipe.net

Tukker, A., Huppes, G., Guinée, J., Heijungs, R., de Koning, A., van Oers, L., et al. (2006). Environmental Impacts of Products (EIPRO) - Analysis of life cycle environmental impacts related to the final consumption of the EU-25. European Commission, Joint Research Centre (DG JRC), Institute for Prespective Technological Studies, Environmental Impacts of Products (EIPRO). Institute for Prespective Technological Studies .

Tukker, A., Poliakov, E., Heijungs, R., Hawkins, T., Neuwahl, F., Rueda-Cantuche, J., et al. (2009). Towards a global multi-regional environmentally extended input-output database. Ecological Economics, 68 (7), 1928-1937.

Resources

EXIOPOL data – www.exiobase.eu

Acknowledgements

The authors would like to acknowledge the whole EXIOPOL team (www.exiobase.eu) for the obvious work in putting together the whole database on what we have based our analysis.

Chapter 18: Structural Decomposition Analysis of the Energy Consumption in China and Russia — An Application of the Eora MRIO Database

Jun Lan and Arunima Malik

Introduction

In this case study we seek to demonstrate the utility of the Eora MRIO database for research and policy applications. We decompose the overall changes in energy consumption over a time period of 21 years (1990 to 2010) according to 6 key determinants: industrial energy intensity; inter-industry structure; final demand structure; final demand destination; population affluence; and population size. Structural Decomposition Analysis (SDA), a comparative-static technique based on input-output analysis, is used to quantify these effects over the entire time period and within four sub-periods. For this purpose, we utilise the input-output tables from the homogeneously-classed version of the Eora MRIO database. China and Russia are chosen as our case study from among the 188 countries available in the Eora database. An important facet of this study is the insights to be gained on the contribution of affluence to an accelerating effect on energy use, whereas energy intensity retarded energy use for China and Russia. The other four determinants had minor impacts.

Background

What is Structural Decomposition Analysis (SDA)?

Structural decomposition analysis (SDA) is an input-output-based technique that breaks down the observed changes in physical variables such as economic, environmental, employment and other social-economic indicators over time into changes in their physical and economic determinants, and quantifies these determinants. The changes in these determinants are understood to be driving the changes in the indicators, either as accelerators or as retardants. In essence, SDA formulates an explained variable as a product of explanatory determinants. This study aims to decompose the energy consumption into six driving determinants (formulated in the equation below), i.e. energy efficiency (**q**), technology (**L**), final demand structure (**u**), final demand destination (**v**), GDP/capita (**y**) and population (**P**), and to quantify the accelerating or retardant effect of each determinant. Fig. 1 illustrates the factors leading to changes in energy consumption and shows the key SDA procedures.

Figure 1: Factors leading to changes in energy consumption

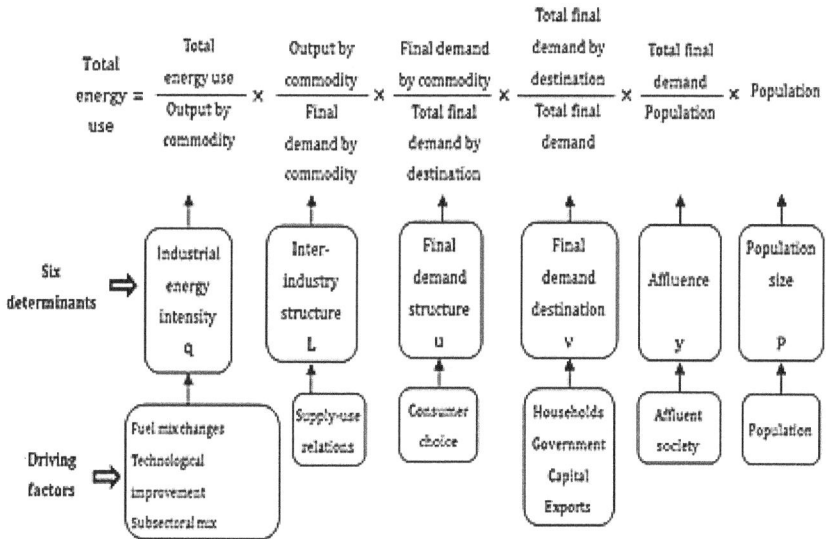

Why Do We Perform SDA?

For decision-makers to respond to environmental issues, it is important to understand the underlying driving forces. SDA is now a widely accepted analytical tool for policymaking on energy issues at a national level. As the determinants of changes in energy use can be quantified by means of a SDA study, it allows the discovery of the real drivers of energy use in production chains and final uses. SDA also supports governments and policy-makers in prioritizing specific sustainability policies. Some examples include policies aimed at

- making industry more energy-efficient by substituting standard electric motors with energy-efficient electric motors (targeting the energy intensity effect),
- improving public transportation to reduce energy use in private cars (targeting the final demand structure effect),
- stimulating some exporting sectors to become more competitive in the international market (targeting the final demand destination effect), and
- controlling the population size of the country such as in China's one-child policy (targeting the population effect).

The value of SDA for decision-making and policy design lies in its ability to uncover the indirect effects that are hidden from many conventional policy appraisals. For example, in Australia there is a spreading awareness that an individual is responsible for the greenhouse gas emissions emitted directly from their cars, their household, or their place of work or study, and the Australian government is actively encouraging citizens to undertake steps to save electricity at home or to reduce their private car use. However, the results of input-output studies on Australian greenhouse gas emissions showed that such government advice only targets 11% of total emissions occurring through direct energy use by households (including the emissions from electricity use, private car use and household use), while 47% of total emissions – caused throughout the economy's production network as a result of household purchases of goods and services – are not addressed at all. Conventional approaches miss the importance of such indirect, hidden effects, and thus fail to benefit from opportunities for efficiently abating energy use and greenhouse gas emissions by reducing goods and services consumption throughout the economy. SDA has the ability to uncover driving forces acting at numerous stages in the economy's supply-chain network.

Why Do We Apply SDA World-Wide?

Applying SDA to global energy consumption provides an effective way to identify which countries and sectors are recording an increase in energy consumption. However, due to the lack of input-output time series tables expressed in constant prices, geographical (country or region) detail, and consistent environmental satellite accounts, much of the prior work is limited to the study of only one country, one national currency unit and the difference between two single years. As far as we are aware, only a few studies deal with structural decomposition analysis of environmental indicators and energy consumption between countries and regions. This may change now that the Eora Multi-Region Input Output (MRIO) database is available to the general public and researchers anywhere are able to apply SDA at the inter-country level and with multiple time intervals.

To illustrate the procedures of SDA, and explain the SDA results for the purposes of this book, we choose China and Russia because of their characteristic and interesting economic development paths.

Data Sources

The Global Multi-Region Input Output database (MRIO) developed by the University of Sydney provides a series of comparable IO tables expressed in a common sector classification and in common monetary units (US$). The conversion of the input-output tables of Eora's 188 countries from the national currency to constant US$ can be achieved by using Purchasing Power Parities (PPP) published by the Organization for Economic Co-operation and Development (OECD). Producer Price Indices (PPIs) published by the U.S. Bureau of Labor Statistics are applied as deflators. For those countries where PPP exchange rates are not available, market exchange rates published by the International Monetary Fund (IMF) can be used.

SDA Results for China and Russia

Decomposition of Changes in Energy Use from 1990 to 2010

Fig.2 shows the contribution of six explanatory factors to changes in energy use (dQ, see Fig. 1) of the Chinese and Russian economies from 1990 to 2010. For China, the affluence effect dy is responsible for 350% of energy use growth within this period. The industrial structure effect dL, final demand structure effect du and population effect dP contribute to a 25%, 15% and 42% increase in energy use, respectively. As a fast developing economy, China has experienced miraculous economic growth, resulting in an increased energy requirement. During the past 21 years, the productive processes in China have become more complex, with longer production chains resulting in an increase of the mechanisation level of the industry with the introduction of more equipment and machines, and ultimately in an increase in embodied energy over all the production stages from mining through manufacturing to distribution.

On the other hand, the intensity effect dq (-160%) and final demand destination dv (-7%) caused a decrease in energy use over the entire period. Owing to an increased attention to energy saving and emissions reduction, the Chinese government has taken effective measures to decrease the energy intensity of China's industries, such as eliminating inefficient enterprises, reducing out-dated capacity, controlling the high-energy-consuming and high-emission industries, promoting ecological compensation, intensifying efforts to reduce energy consumption in high-energy-consuming industries, implementing differential power pricing policy and introducing resource tax reform. To sum up, during these 21 years, accelerating effects are significantly more prevalent than retarding effects, therefore, energy use in China increased by nearly 40,000 PJ.

For Russia, energy intensity dq (-104%) has significantly more retarding effects compared to accelerating effects, causing the decrease of energy use in Russia by around 7500PJ in the past 21 years. Similar to China, the affluence effect dy plays an important role in energy use growth (80%). Among other factors, industrial structure effect dL and final demand destination effect dv worked as accelerators (with 6% and 0.3% respectively), while final demand structure effect du and population effect dP acted as retardants (with -0.2% and -4% respectively).

Figure 2: Contribution of six effects to changes in energy use of China and Russia from 1990 to 2010

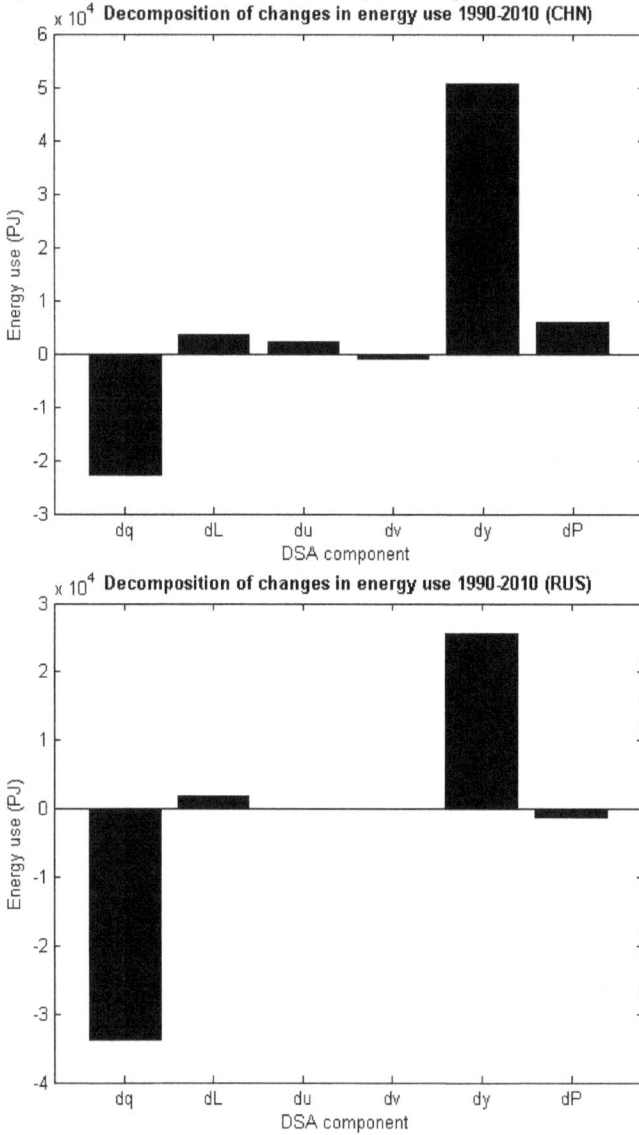

Macro-results- China — 20-year average

Accelerators:

		Retardants:	
Industrial structure	25.3%	Energy intensity of industry	-157.7%
Final demand structure	15.5%		
Population affluence	353.0%	Final demand categories	- 6.8%
Population growth	41.6%		

+ 435.5 %   **- 164.5 %**

+ 271 %

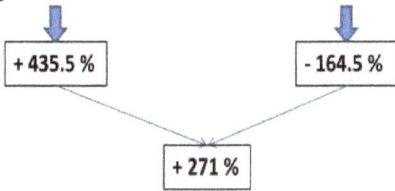

Macro-results- Russia — 20-year average

Accelerators:

		Retardants:	
Industrial structure	5.5%	Energy intensity of industry	-104.2%
Final demand categories	0.3%	Final demand structure	- 0.2%
Population affluence	79.0%	Population decrease	- 3.7%

+ 84.8 %   **- 108.1 %**

- 23.3 %

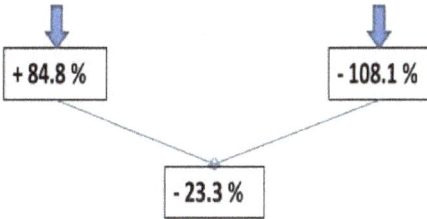

The results for China and Russia for the whole period shown above give a general idea about what happened in the last 21 years. They do not show, for example, the changes due to the influence of short-term events, such as the political measures of the Chinese government to open the special economic zone for the international market since 1990, the proposal for an 'energy saving and emission reducing' policy in 2005, the 2008 Beijing Olympic Games, the Expo 2010 Shanghai China, the collapse of the Former Soviet Union in 1991, the financial crisis of Russia in 1998, and the Russian economic resurgence since 2000. In order to obtain a better reflection of these events, the period of 21 years was divided into sub-periods of five-years. In the following section, the results of SDA for all four sub-periods are demonstrated.

Decomposition of Changes in Energy Use in Four Sub-Periods

Fig.3 illustrates the contribution of all effects to changes in the energy use for China and Russia and the development of these changes (sum over all effects) in the four sub-periods between 1990 and 2010. It can be observed that within 21 years, energy use in both these countries increased because of the accelerating effects of changes in energy use, except a slight decrease in the sub-period of 2005-2010 in China.

Figure 3: Contribution of all effects to changes in China's and Russia's energy use from 1990 to 2010

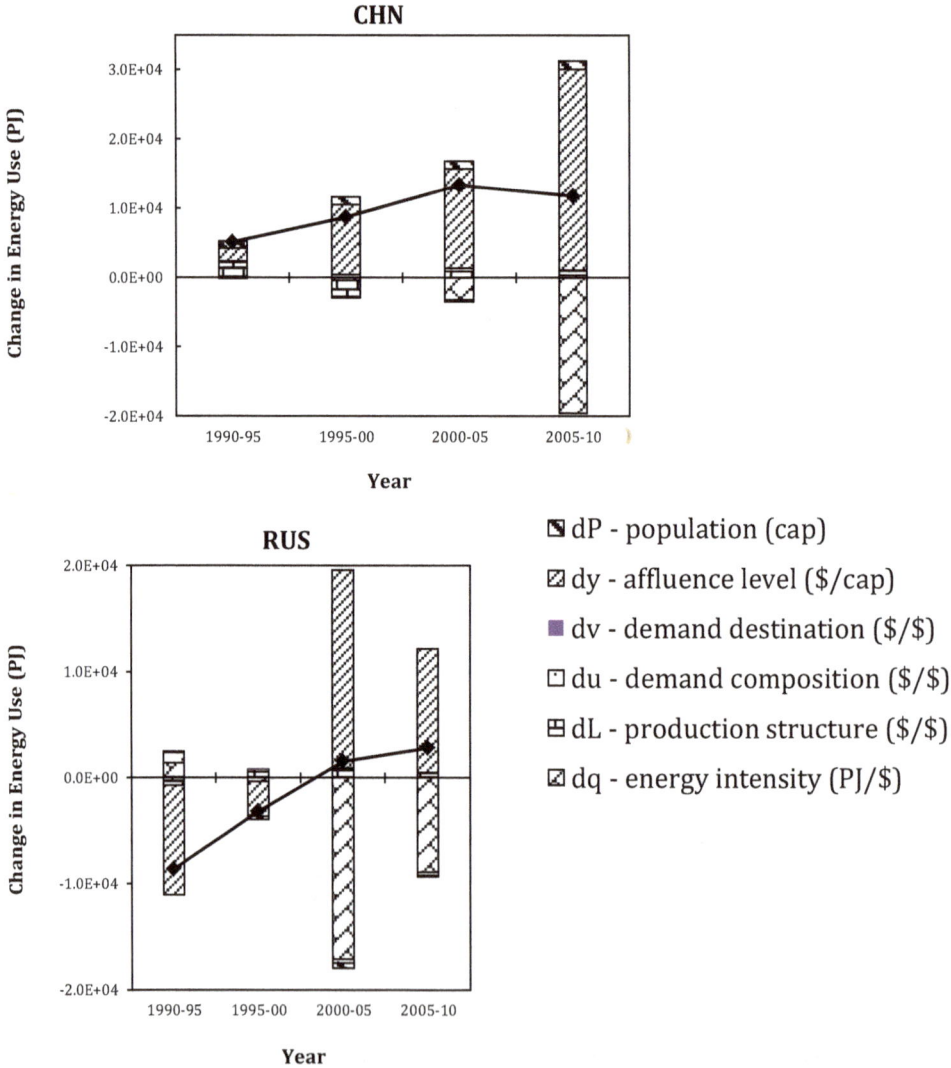

For China, the division of the intensity effect dq (changes in energy use due to changes in direct energy intensity of industries) in the sub-periods tells a similar story to that of the SDA results for the total period (1990-2010), except dq contributed to a slight accelerating effect only in the first period. In the last three periods it tended to be retarding and became the most retarding effect from 2000 to 2010. This phenomenon means that the sectors of China's economy became more efficient in their energy use over time, due to continued efforts to implement the 'energy saving and emission reducing' policy and levying a resource tax across the whole economy. The industrial structure effect dL (changes in energy use due to changes in inter-sectoral dependencies) showed positive trends in the first, third and fourth periods, but retarding effects in the second period. As the sectors recorded a retarding trend for dq, the retarding effect of dL in 1995 to 2000 can be explained by inter-sectoral changes in favour of less energy –intensive sectors. From 2000 onwards, the effect of dL became insignificant. During the entire period, final demand structure effect du and final demand destination effect dv exerted no significant contribution to changes in energy use. The changes in the level of economic activity dy – measured in GDP per capita – always contributed to an increase in the total energy use. It can be observed that the evolution of dy and total energy consumption dQ proceeded in a similar way, even dQ showed a slight decreasing trend from 2005 to 2010. This evolution can be explained by a miraculous increase in the Chinese GDP since 1995 as a consequence of some activities of economic stimulus and the installation of more energy-efficient projects. As expected, the population effect dP over energy use continued to accelerate in each period, as it has in the past.

The results of the SDA show that changes in the energy use in China from 1990 to 2010 were mainly influenced by changes in affluence, direct energy intensity and population, rather than changes in inter-sectoral dependencies, final demand structure and final demand destination. Considering the results of the SDA study on China, the major accelerating contribution (dy) is not a suitable target that the Chinese government can take actions to bring about a reduction in energy use. Economic development is important and desirable in the view of Chinese decision-makers, mainly to achieve a more equitable income distribution, which in turn influences directly and indirectly China's total energy use. Therefore, policies should be introduced aiming at the intensity effect dq and the population effect dP. The Chinese government will continue to enforce and enhance the existing energy-saving measures on the sectors with major energy requirements, with high final demand in monetary terms and with high energy intensities, and control the population growth rate at lower annual rates.

The results of our SDA study for Russia differ markedly from those for China. The evolution of the intensity effect dq proceeded as an accelerator in the first period and as a retardant in the following three periods. Because of its continuous retarding contribution to total energy consumption dQ, dq became the most important effect over the total period. This evolution means that only since 2000, the sectors of the Russian economy have become more efficient in their energy use, and relatively inefficient technology prevailed for nearly 10 years after the FSU collapse. As shown in Fig.3, the affluence effect dy played the second-most significant role in the changes to energy use. dy contributed to a decrease in total energy use in the first two periods, but was responsible for an increase of energy consumption dQ in the last two periods.. The negative effects of dy from 1990 to 2000 clearly highlight the aftermath of the collapse of the Former Soviet Union in 1991. The high participation of capital expenditure in GDP usually means an increase of energy requirements because of its implications for construction. This can be explained by the fact that from 2000 the Russian economy started to recover from the financial crisis in 1998. As the retarding effect, caused by the substitution of inputs to the production of certain products by more energy-intensive inputs, was offset by the accelerating effect induced by increasing energy requirements from fast economic growth, overall energy consumption dQ continued to grow. In order to control the growing energy consumption, the Russian government can apply measures to reduce the energy consumption of industries with high energy requirements and with high final demand in monetary terms (as in the case of construction).

Conclusions

The results of this SDA study illustrate an increase in energy demand in both China and Russia, even after the implementation of effective measures to progress the current economy towards sustainability. In summary, affluence contributed an accelerating effect on energy use, while energy intensity retarded energy use. China has experienced massive growth in economy since the 1990s, facilitated by many milestones. Since 1990, the Chinese government decided to open Shanghai and Shenzhen economic zone for the international market and established the Shanghai and Shenzhen Stock Exchange offices. In order to accelerate the infrastructure development and improve the ecological environment in the western region, the Chinese government started to carry out the grand western development program since 2000. In 2001, China became the 143rd member of the World Trade Organization (WTO).The 2008 Beijing Olympic Games and the Expo 2010 Shanghai China further motivated the expansion of domestic investment and consumption demand. Under the stimulus of these economic activities, China has experienced an economic growth miracle since 1995. Two influential economic activities for Russia are the collapse of the Former Soviet Union in 1991 and the economic resurgence since 2000. These historic events explain the reasons for the growth in affluence, and the driving factors for the increase in energy use. The installation of more energy-efficient projects was beneficial and the total energy consumption increased at a stable but comparatively low speed.

Governments can use SDA results to gain an understanding of the importance of applying pertinent policies to the changes in energy use caused by affluence, population and industrial energy intensity. A decline in energy use can be expected as China and Russia move from a heavy-industries economy towards a services-oriented economy, but the actual effects will depend on the construction of "green industries", and on the provision of effective incentives to industries with both energy-efficient and non-energy-intensive characteristics. It is worth pointing out that although the SDA results provide basic directions for policy design, it is important to carefully assess the policies aimed at reducing energy use before putting them into force to avoid exacerbating the energy shortage problem.

Further reading

Dietzenbacher, E., & Los, B. (1998). Structural Decomposition Techniques: Sense and Sensitivity. *Economic Systems Research, 10*, 307-324.

Hoekstra, R., & van den Bergh, J. (2002). Structural decomposition analysis of physical flows in the economy. *Environmental and Resource Economics, 23*, 357-378.

Kagawa, S., & Inamura, H. (2001). A structural decomposition of energy consumption based on a rectangular input-output framework: Japan's case. *Economic Systems Research, 13*, 339-363.

Lenzen, M. (2006). Structural Decomposition Analysis and the Mean-Rate-of-Change Index. *Applied Energy, 83*, 185-198.

Lenzen, M., Geschke, A., Kanemoto, K., & Moran, D. (2011). *The Eora global multi-region input-output tables.* Retrieved 2012 from World MRIO: http://www.worldmrio.com

Pardey, P., Roseboom, J., & Craig, B. (1992). A Yardstick for International Comparisons: An Application to National Agricultural Research Expenditures. *Economic Development and Cultural Change, 40*, 333-349.

Rose, A. (1999). Input-output structural decomposition analysis of energy and the environment. In J. C. J. M. van den Bergh (Ed.), *Handbook of Environmental and Resource Economics* (pp. 1165-1179). Cheltenham, UK: Edward Elgar.

Rose, A., & Casler, S. (1996). Input-output structural decomposition analysis: a critical appraisal. *Economic Systems Research, 8*, 33-62.

Viet, V. (2002, October 21-25). The role of the supply and use tables in double-deflation method. *UNESCAP/ECO/UNSD/SIS Subregional Workshop on Implementation of the 1993 System of National Accounts.* Ankara, Turkey.

Wachsmann, U., Wood, R., Lenzen, M., & Schaeffer, R. (2009). Structural decomposition of energy use in Brazil from 1970 to 1996. *Applied Energy, 86*, 578-587.

Wier, M. (1998). Sources of changes in emissions from energy: a structural decomposition analysis. *Economic Systems Research, 10*, 99-112.

Wood, R. (2009). Structural decomposition analysis of Australia's greenhouse gas emissions. *Energy Policy, 37*, 4943-4948.

Acknowledgements

The writing of this chapter was supported by the Australian Research Council through Discovery Projects DP0985522 and DP130101293.

Chapter 19: Estimating Global Environmental Impacts of Goods and Services Produced in Japan Using a Global Link Input–Output Model (GLIO)

Keisuke Nansai, Shigemi Kagawa, Yasushi Kondo, Susumu Tohno,
and Sangwon Suh

Toward Building a Japanese Input–Output Life Cycle Assessment Database with a Global System Boundary

This chapter introduces a coefficient called "intensity of embodied global environmental burden", which is conveniently used for the life cycle assessment (LCA) of goods produced and services provided in Japan, by application of the global link input–output model, a simplified version of the MRIO model, the structure of which is described in Chapter 8: 'Simplification of multiregional input–output structure with a global system boundary: The global link input–output model'.

The increasing variety of imported products that we encounter in our everyday lives bears testament to the fact that international trade is expanding. Consequently, when attempting to manage environmental emissions through the life cycles of products, it has become increasingly important to ascertain the environmental impacts that are generated indirectly in foreign countries through the global supply chain. Particularly, the spread of carbon footprint labels and the introduction of new calculation and reporting standards has strengthened the control of supply chain emissions of greenhouse gases (GHG). For instance, the Scope 3 of the Greenhouse Gas Protocol (GHG Protocol) initiative and the International Organization for Standardization (ISO) each provide guidelines on the quantification and reporting of greenhouse gas emissions and removals at the organization level. Generally, supply chains associated with commodity production and corporate business activities are no longer limited to enterprises within a domestic economy. Rather, such chains now extend throughout the world. Consequently, elucidating the specific emissions associated with particular production processes in a global context requires considerable time and labor costs.

One way to produce a comprehensive account of emissions is to build an LCA database that is useful to estimate the global environmental impacts associated with the production of goods and provision of services. Two approaches are currently used to construct LCA databases: process-based LCA and input–output LCA (IO-LCA). Depending on data availability, process-based LCAs allow for the sequential inclusion of data related to the environmental burdens associated with imports in a supply chain. However, guaranteeing the completeness of global supply chain descriptions using process-based LCA is often difficult. Moreover, the method might fail to detect key emission sources in the global supply chain. In contrast, IO-LCA theoretically guarantees the completeness of global supply chain descriptions by adopting the framework of a multiregional input–output (MRIO) model. IO-LCA with MRIO is therefore regarded as better suited for identifying global supply chains with high environmental burdens and for prioritizing processes for which inventory data must be elaborated.

We compiled an IO-LCA database for Japanese products using a global link input–output (GLIO) model (Nansai et al., 2009, 2012a). A GLIO model was employed to reduce the time and labor associated with data processing, while consistently and adequately describing the relations between the Japanese economy and the global supply chain. Fundamentally, a GLIO model is a simplified MRIO model that is tailored to a specific purpose or type of analysis. For additional information related to how a GLIO model can be simplified relative to a conventional MRIO model, please refer to chapter 8.

We used a GLIO model to estimate the GHG emissions generated in countries around the world that are associated with one unit of commodity production in Japan. Specifically, we examined goods and services of 406 types defined in the 2005 IO tables for Japan. We referred to the GHGs generated around the world per unit of production as the "embodied global-GHG intensity". It is noteworthy that, in this study, one unit of commodity production is equivalent to one million yen (M-JPY). For the GLIO model, we set the number of countries and regions used in the model to 231, including Japan, and calculated the embodied global GHG intensity of Japan using an environmentally extended GLIO table for 2005. For an explanation of how the intensity was determined from the GLIO table with the GLIO model, please see the earlier literature (Nansai et al., 2012b).

What is an Embodied Global GHG Intensity?

Figure 1 shows that the embodied global GHG intensity can be represented conceptually. Using the example of automobile production in Japan, GHGs are emitted on site as energy is consumed during assembly at automobile production facilities. We refer to these direct emissions in Japan at the facilities as "category D" emissions. Various automobile components are procured from materials and equipment manufacturers in Japan, and GHGs are emitted during the production of these components. We refer to these emissions in Japan through domestic (Japanese) supply chains as "category S" emissions.

Imported components and raw materials are also used by materials and automobile equipment manufacturers in Japan. This direct and indirect use of imported products induces production abroad, which in turn is associated with the emission of GHGs. We categorize emissions abroad from *foreign supply chains* as "category F" emissions.

An embodied global-GHG intensity is therefore the sum of GHG emissions emitted in each section of (D), (S) and (F) that are caused during the production of one M-JPY worth of automobiles in Japan. In other words, the embodied global-GHG intensity means the GHGs emitted either directly or indirectly around the world during each stage of a product's life cycle (e.g. during raw material extraction, processing and assembly). By calculating this intensity for the goods and services of 406 types, we can compile an IO-LCA database that incorporates information for all global supply chains. This database is expected to facilitate the calculation of carbon footprints of products and services and GHG emission management based on the Scope 3 corporate standard.

Figure 1: Global system boundary of GHG emissions incorporated in an embodied global-GHG intensity for a Japanese product in a global link input–output model.

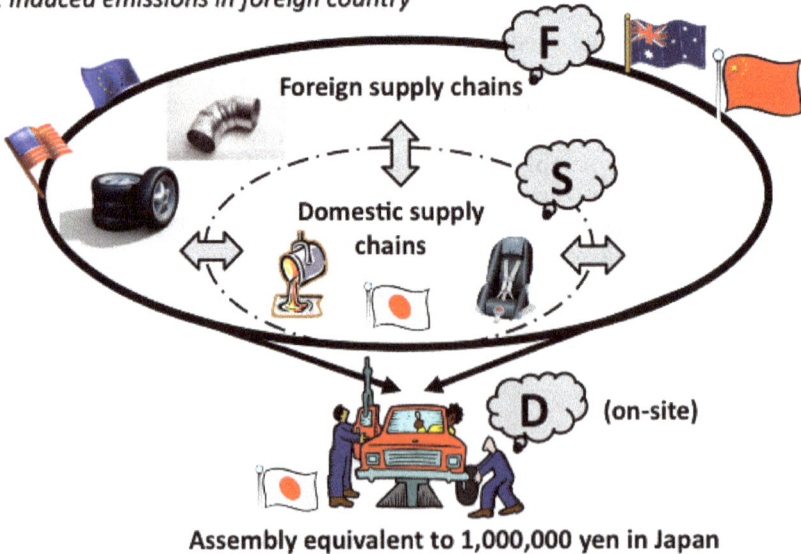

Embodied global-GHG intensity [CO_2eq/M-JPY] = D + S + F

D: Direct emissions in Japan
S: Induced emissions in Japan
F: Induced emissions in foreign country

Foreign supply chains

Domestic supply chains

D **(on-site)**

Assembly equivalent to 1,000,000 yen in Japan

Japanese Products with Embodied Global-GHG Intensities

Here we characterise Japanese goods and services with regard to their embodied global-GHG intensities. First, we will examine products with high embodied global-GHG intensities. Table 1 presents sectors with the ten highest embodied global-GHG intensities in Japan. The respective contributions of (D), (S) and (F) show phases during which GHG emissions occur in a given commodity's supply chain: (D) Direct emissions in Japan during production; (S) Induced emissions in Japan through the supply chain; and (F) Induced emissions abroad in foreign supply chains.

The sector with the highest embodied global GHG intensity by far is *Cement* followed by *Pig iron*, *On-site power generation* and *Crude steel (converters)*. These products are the only sectors with intensities exceeding 30 t-CO$_2$eq/M-JPY. In other words, this means that much of the carbon footprint per million-yen production of products in Japan does not exceed this value. In this case, the value 30 t-CO$_2$eq/M-JPY can be assumed as the upper limit when verifying the propriety of the result of carbon footprint estimation using the process-based LCA. The same conception is applicable to the calculation of carbon footprint associated with corporate production activities. We assume that a company produces Commodity A in the annual amount of p (M-JPY) and Commodity B in the annual amount of q (M-JPY). If the carbon footprint per unit of production of Commodity A and Commodity B (F_a and F_b) does not exceed 30 t-CO$_2$eq/M-JPY, the weighted average carbon footprint of producing A and B ($F_a \times p + F_b \times q/(p+q)$) also will not exceed 30 t-CO$_2$eq/M-JPY. This value can be therefore useful as a standard for the upper limit of carbon footprint per unit of corporate activities.

Table 1: Japanese products with the ten highest embodied global-GHG intensities, as well as the proportions of direct and induced emissions associated with production in Japan and abroad.

Rank	Sector number and name	Embodied global-GHG intensity	(D) Share of direct emissions in Japan	(S) Share of induced emissions in Japan	(F) Share of induced emissions in foreign country
		[t-CO$_2$eq/M-JPY]	[%]	[%]	[%]
1	Cement	138	91.6	6.5	2.0
2	Pig iron	72.6	84.3	5.6	10.1
3	On-site power generation	68.8	92.2	3.1	4.7
4	Crude steel (converters)	45.5	5.6	81.6	12.8
5	Electricity	29.1	85.3	6.2	8.5
6	Ocean transport	27.3	52.0	2.5	45.6
7	Ready-mixed concrete	27.3	1.0	94.9	4.1
8	Hot rolled steel	26.8	4.3	79.9	15.8
9	Coal products	21.5	40.0	5.9	54.1
10	Industrial soda chemicals	21.2	17.7	66.1	16.3

Excluding the 6th rank "Ocean Transport" and 10th rank "Industrial Soda Chemicals," the rest are sectors related to steel, cement, and electric power. This suggests that reduced expenses for steel and cement products in manufacturing, corporate activities, and infrastructure development would not only contribute to lower cost of production, but help effectively control GHG emissions. In addition, observing the shares of these sectors in the emission categories (D), (S), and (F) facilitates understanding of whether they contribute to reduction of domestic or overseas emissions. This information is likely to be extremely beneficial for elucidating how companies fulfill their social responsibilities of reducing GHG emissions in the whole world.

The intensities of "Ocean Transport" and the 9th rank "Coal Products," for instance, contain 46% and 54%, respectively, of Category (F). Whereas the values of these GHG intensities are the same level as those of the 7th rank "Ready-mixed Concrete" and "Industrial Soda Chemicals," the expenses for these products are equally related to both domestic and overseas emissions, and reduction of such expenses is characterized by its contribution to a decrease in overseas emissions.

Regarding the contribution of foreign emissions to embodied global-GHG intensities, Table 2 presents those commodities with the highest associated category (F) emissions, i.e. induced emissions in foreign countries. At the top of the list, with an embodied global-GHG intensity of 86%, is *Rolled and drawn aluminum*, indicating that the production of aluminum products in Japan induces substantial emissions abroad. This sector is followed by *Other non-ferrous metal products* at 83%, *Animal feed* at 82%, *Vegetable oils and meal* at 76%, and *Other non-ferrous metals* at 76%. These sectors also engender considerable emissions in global supply chains through the consumption of metal resources and crops used in the production of animal feed and forage.

As exemplified by the current emission control framework of the United Nations Framework Convention on Climate Change (UNFCCC), these sectors having a high percentage of overseas emissions would not offer much incentive for reduced use when only the GHG emissions in Japan are specifically examined. By expanding supply chains globally, however, even commodities such as those presented in Table 2 would clearly contribute to substantial reduction of GHG emissions, which motivates the control of the amount of use and expenses. We hope that the use of embodied global-GHG intensities will provide businesses, governments, and consumers with an opportunity to reconsider new ways to reduce their GHG emissions.

Table 2: Japanese products with the ten highest proportions of induced foreign emissions in their embodied global-GHG intensities

Rank	Sector number and name	Share of foreign emissions in embodied global-GHG intensity [%]
1	Rolled and drawn aluminum	86
2	Other non-ferrous metal products	83
3	Feeds	82
4	Vegetable oils and meal	76
5	Other non-ferrous metals	76
6	Copper	75
7	Processed meat products	75
8	Nuclear fuels	75
9	Flour and other grain mill products	72
10	Gas supply	71

List of Embodied Global-GHG Intensities and Global Intensities of Other Environmental Burdens

We have calculated the embodied global-GHG intensities in Table 1 for 406 sectors of goods and services in Japan and provide them on the website (http://dx.doi.org/10.1021/es2043257) free of charge as the Supporting Information (SI) attached to Nansai et al. (2012b). The data are available as a list in PDF format and as tab separated values in the text format. The text file can be dragged and dropped to a Microsoft Excel ® worksheet to create a list of the intensities in the same style as the list in the PDF file.

Table 3: Japanese products with the three highest intensities of each environmental burden and their proportions of direct and induced emissions associated with production in Japan and abroad.

	Intensity	(D) [%]	(S)[%]	(F)[%]
Energy consumption	[GJ-NCV/M-JPY]			
1 On-site power generation	802	91.8	3.6	4.5
2 Pig iron	729	82.2	7.5	10.4
3 Cement	600	76.6	19.3	4.2
CO$_2$	[CO$_2$/M-JPY]			
1 Cement	135	92.1	6.5	1.4
2 Pig iron	71.1	85.8	5.6	8.6
3 On-site power generation	67.5	93.4	3.1	3.5
CH$_4$	[CO$_2$eq/M-JPY]			
1 Beef cattle	8.85	51.2	33.5	15.3
2 Dairy cattle farming	6.88	82.3	1.3	16.4
3 Slaughtering and meat processing	4.23	0.0	72.1	27.9
N$_2$O	[CO$_2$eq/M-JPY]			
1 Crops for beverages	9.08	98.8	0.7	0.6
2 Crops for feed and forage	7.08	95.1	1.2	3.7
3 Fowls and broilers	4.66	68.1	1.2	30.7
HFCs	[CO$_2$eq/M-JPY]			
1 Refrigerators and air conditioning apparatus	3.57	84.0	10.0	6.0
2 Other resins	0.926	97.6	0.3	2.2
3 Cosmetics, toilet preparations and dentifrices	0.623	96.4	2.3	1.2
PFCs	[CO$_2$eq/M-JPY]			
1 Semiconductor devices	4.37	99.7	0.1	0.2
2 Liquid crystal element	0.424	73.3	19.5	7.2
3 Integrated circuits	0.371	84.2	11.1	4.7
SF$_6$	[CO$_2$eq/M-JPY]			
1 Semiconductor devices	1.07	98.5	0.3	1.2
2 Non-ferrous metal castings and forgings	0.960	98.3	1.3	0.4
3 Liquid crystal element	0.581	87.3	3.6	9.1
NO$_x$	[kg/M-JPY]			
1 Cement	149	88.6	8.3	3.1
2 Ocean transport	125	21.0	1.6	77.4
3 Marine fisheries	95.8	94.4	2.5	3.1
SO$_x$	[kg/M-JPY]			
1 Cement	130	87.0	7.9	5.1
2 Marine fisheries	115	94.3	2.1	3.6
3 Ocean transport	109	11.5	1.3	87.2

The GHG intensities are also organized based on their types (carbon dioxide (CO$_2$), methane (CH$_4$), nitrous oxide (N$_2$O), perfluorocarbons (PFCs), hydrofluorocarbons (HFCs), and sulfur hexafluoride (SF_6)). In addition to GHGs, the intensities of energy consumption on a net calorific value (NCV) basis, nitrogen oxides (NO_x), and sulfur oxide (SO_x) were calculated and included in the SI above. Table 3 presents sectors with the three highest intensities of each environmental load substance. CO_2 comprises approximately 90% of Japan's GHG emissions, resulting in the same top sectors of CO_2 as those of GHGs. Like CO_2, energy consumption, NO_x, and SO_x are environmental burdens associated with fossil fuel consumption, which places the sectors also presented in Table 1 at the top. Such substances as CH_4 and N_2O show a large proportion of agricultural intensity and HFCs, PFCs, and SF_6 are linked largely to the intensity of specific industrial products. These results thereby reveal the types of goods and services that contribute to an increase or decrease in environmental burdens depending on the types of the burdens. Using databases to confirm the types of environmental burdens that materials and components used by companies and consumables carry globally, we must consider what each of us can do to support global environmental management.

Implications of the Embodied Global-GHG Intensities for Use in LCA

If an LCA practitioner knows the 2005 producer price (M-JPY) of a commodity, then the global GHG emission (t-CO_2eq) of the commodity can be estimated solely by multiplying that price by the embodied global-GHG intensity (t-CO_2eq/M-JPY). Although this is straightforward, the accuracy of the price or physical amounts of these commodities can have a significant bearing on the life cycle emissions calculated, and consequently, on the overall reliability of the result. This holds true particularly for IO-LCA and hybrid LCA of manufacturing processes and technologies with inputs consisting of high embodied global-GHG intensity commodities, such as those presented in Table 1. The LCA practitioner should exercise particular caution when quantifying prices and physical quantities of such commodities, assigning due consideration to year-on-year differences in prices and/or technologies.

In process LCA, commodities with high global-GHG intensity should be identified at the screening phase because they deserve particular attention with respect to the collection of detailed process data. For most such commodities, it is the direct emissions that are the main contributor to the high embodied global-GHG intensity. The on-site emissions occurring at the production site of the commodity should therefore be afforded greater consideration than the emissions associated with that commodity's global supply chain.

A commodity with a high proportion of overseas emissions generates large GHG emissions in the overseas supply chain associated with the imported goods and services used in the domestic supply chain. For commodities with large shares of foreign emissions, such as those presented in Table 2, the LCA practitioner should prioritize the collection of process and emission data for the imports in the supply chain. In process LCA, because commodities with large domestic emissions (both direct and indirect) tend also to have large foreign emissions, due consideration should be devoted to identifying the foreign emissions associated with such commodities.

However, data related to processes and emissions in foreign countries are often scarce and costly to collect. When conducting LCA on such commodities, applying the IO-LCA and hybrid LCA with the embodied global emission intensities as described in this chapter can nonetheless have major benefits over process LCA, which would generally require that the practitioner ignore the contribution of foreign emissions associated with the production of a commodity because of cost and data availability constraints.

Application: IO-LCA for the Whole of Japan

The IO tables for Japan show the annual monetary demand for goods and services induced by the final demand sectors in Japan. Multiplying this monetary demand for each product by the embodied global-GHG intensity yields an estimate of the GHG emissions generated in Japan and abroad that were induced by Japanese domestic final demand. GHG emissions in Japan and abroad are synonymous with the carbon footprint and consumption-based emissions for the whole of Japan. To estimate Japan's embodied GHG emissions (i.e., national carbon footprint), we used the Japanese domestic final monetary demand for each product for 2005 described in the 2005 Japanese input-output tables and multiplied it by the embodied global-GHG intensity corresponding to the product. In this way, Japan's total carbon footprint was estimated as 1,675 Mt (megatons) CO_2eq.

The structure of Japan's carbon footprint is shown in Figure 2. The outer ring of the pie chart shows the contribution of domestic final demand to Japan's carbon footprint, broken down into five categories (1. Household consumption, 2. Government consumption, 3. Public fixed capital, 4. Private fixed capital and 5. others). The inner ring of the chart provides a breakdown of the emissions of each category of final demand into [a] direct domestic emissions, [b] induced domestic emissions from the supply chains of commodities demanded, and [c] induced overseas emissions from the global supply chains of commodities demanded. This figure depicts that household consumption accounts for the largest share of emissions: 1028 Mt CO_2eq (61%). The domestic emissions are a combination of 184 Mt CO_2eq (11%) from the direct combustion of fuels such as gasoline and kerosene and 515 Mt CO_2eq (31%) emitted in the domestic supply chain underlying production of consumed commodities. The overseas emissions, 329 Mt CO_2eq, represent 20% of total emissions, which implies the degree to which global GHG emissions are controlled by our daily life activities.

Outside the country, the Japanese economy generated global emissions of 541 Mt CO_2eq. Figure 3 presents a schematic representation of the global distribution of Japan's carbon footprint. The largest contributor was China, which accounted for 165 Mt CO_2eq, or 30% of these global emissions. Next were the US (12%), Australia (6.5%), Saudi Arabia (4.8%), Russia (4.2%), with the top ten rounded out by Indonesia (3.1%), the UAE (2.7%), Canada (2.6%), South Korea (2.5%), and Thailand (1.8%). In an effort to reduce its carbon footprint, the role that Japan should play in international GHG emission control is to reduce emissions abroad through joint development of low-carbon technologies and technology transfer with countries that induce such emissions in addition to controlling its domestic emissions.

Figure 2: Contributions of five final demand categories to Japan's carbon footprint in 2005 and the composition of each category's emission locations: in Japan (direct), in Japan (supply chain) or overseas.

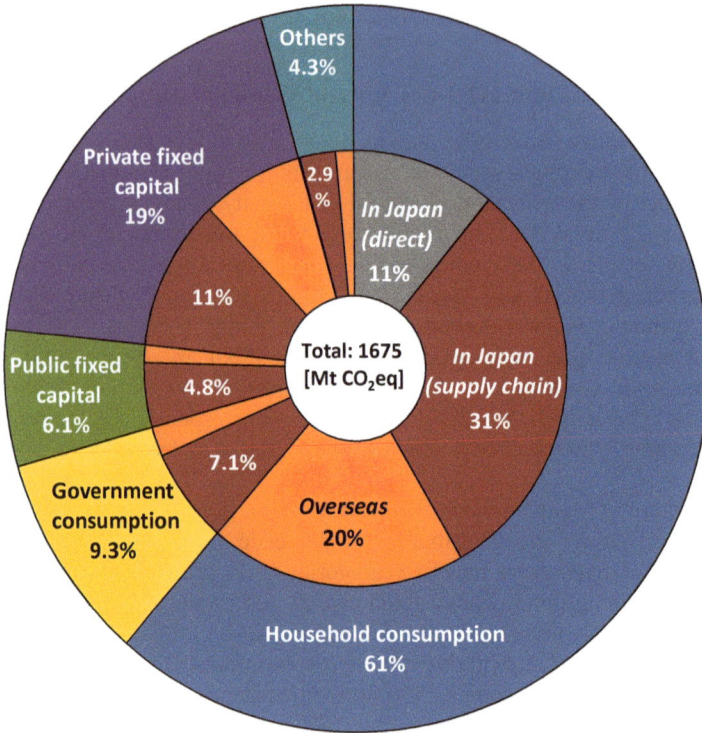

Figure 3: Global distribution of the Japanese carbon footprint (GHG emissions) in 2005 (Unit: Mt CO_2eq).

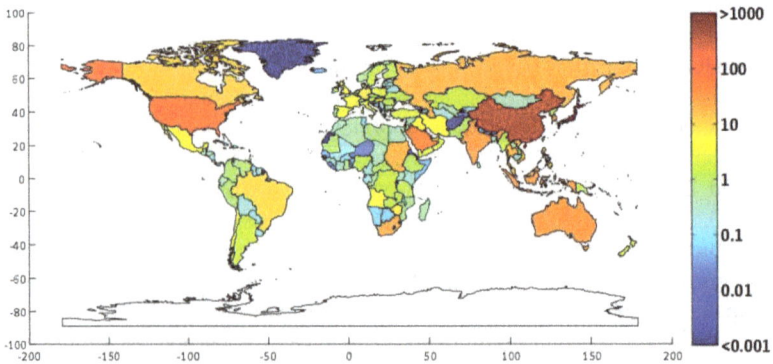

Further reading

Davis, S., & Caldeira, K. (2010). Consumption-based accounting of CO2 emissions. *Proceedings of the National Academy of Sciences, 107* (12), 5687-5692.

Hertwich, E., & Peters, G. (2009). Carbon Footprint of Nations: A Global, Trade-Linked Analysis. *environmental Science and Technology, 43*, 6414-6420.

Nansai, K., Kagawa, S., Kondo, Y., Suh, S., Inaba, R., & Nakajima, K. (2009). Improving the Completeness of Product Carbon Footprints Using a Global Link Input-Output Model: The Case of Japan. *Economic Systems Research, 21* (3), 267-290.

Nansai, K., Kagawa, S., Kondo, Y., Suh, S., Nakajima, K., Inaba, R., et al. (2012). Characterization of economic requirements for a "carbon-debt-free country". *Environmental Science & Technology, 46* (1), 155-163.

Nansai, K., Kondo, Y., Kagawa, S., Suh, S., Nakajima, K., Inaba, R., et al. (2012). Estimates of Embodied Global Energy and Air-Emissions Intensities of Japanese Products for Building a Japanese Input-Ouput Life Cycle Assessment Database with a Global System Boundary. *Environmental Science & Technology, 46* (16), 9146-9154.

Suh, S. (2004). Functions, commodities and environmental impacts in an ecological-economic model. *Ecological Economics, 48*, 451-467.

Suh, S., Lenzen, M., Treloar, G., Hondo, H., Horvath, A., Huppes, G., et al. (2004). Systems boundary election in life-cycle inventories usying hybrid approaches. *Environmental Science and Technology, 38* (3), 657-664.

Resources

ISO 14040:2006, Environmental management -- Life cycle assessment -- Principles and framework

Acknowledgements

This research was partially supported by a Grant-in-Aid for Scientific Research in Innovative Areas (KAKENHI: Grant No. 4003-20120005) from the Ministry of Education, Culture, Sports, Science and Technology, Japan.

Chapter 20: Carbon Emissions and Materials Embodied in Emerging Economies' Trade: An Application of the Global Resource Accounting Model (GRAM)

Martin Bruckner, Leisa Burrell, Stefan Giljum, Christian Lutz, and Kirsten S. Wiebe

Background

A number of recent reports on global environmental developments illustrate worrying trends that are closely linked to the global industrialisation process and have led to a significant increase in resource use and corresponding negative environmental impacts.

Both global material use and carbon dioxide emissions have increased by more than 50% since 1980, causing some of the most pressing environmental challenges of our time. These include climate change, changes to land cover and use, exerting pressure on biodiversity and arable land, as well as the production of solid wastes and air and water pollution.

Emerging economies play an increasingly important role in determining these global environmental trends. Within only two decades the share of the 16 largest emerging economies in global material extraction increased from 25% to nearly 40%. Similarly, growth rates in carbon emissions were highest for emerging economies, most notably China. This increase in environmental pressures in emerging economies is closely linked to changes in the global system of production and trade. Apart from a temporal decrease in 2009 due to the global economic crisis, trade volumes have been rapidly increasing in the past years, at growth rates constantly higher than growth rates in world production. This is because globalisation has led to an increased international division of production and labour, with some world regions specialising in resource extraction, others in processing and manufacturing activities and final consumption taking place yet on another continent.

The increasing spatial separation of the extractive and manufacturing industries from the final consumption of goods and services raises the question of responsibility for the increased environmental pressures occurring in countries with expanding rates of industrialisation. At present, environmental accounting systems such as that applied under the United Nations Framework Convention on Climate Change focus on a territorial accounting approach. This accounts for production-based greenhouse gas emissions and material extraction within a clearly defined territory, usually national boundaries. Such an accounting approach places responsibility for the environmental impacts stemming from industry firmly with the producer country. In contrast, consumption-based accounting considers all emissions emitted or materials used along global production chains of goods and allocates these environmental factors to the final consumer. Research that examines the environmental inputs embodied in products is therefore an essential foundation that will help inform scientific and policy discussions regarding consumer responsibility.

For some decades, environmentally extended input-output models have been used to investigate environmental inputs embodied in final demand, such as carbon emissions, land and water embodied in goods and services. The input-output approach clearly defines system boundaries by including the whole economy, and therefore allowing for a consistent analysis of global shifts of factor embodiments via trade. Moreover, multi-region models consider country specific technologies and economic structures within one model by linking domestic input-output tables via international trade.

In this chapter we employ the Global Resource Accounting Model (GRAM), an environmentally extended multi-regional input-output model which covers the whole world economy divided into 53 countries and two regions, complemented by data on the global extraction of materials from the SERI Material Flows Database and with data from the International Energy Agency on global emissions of carbon dioxide from fossil fuel combustion. For a detailed description of the GRAM model, please refer to Chapter 9 of this book. We also analyse the environmental burden from a consumption perspective, that is, investigating consumer responsibility, taking into account the global flows of embodied materials and embodied carbon dioxide emissions from international trade. In particular, we analyse six major emerging economies, the BRICSA countries (Brazil, Russia, India, China, South Africa, and Argentina).

Key questions that we address are as follows:

- How has the virtual relocation of carbon emissions and materials from global trade changed between 1995 and 2005?
- What part do the major emerging economies play in these developments?

- Do materials and carbon emissions embodied in international trade follow similar or divergent patterns?
- What is the role of highly industrialised countries and how do environmental indicators change when switching from producer to consumer responsibility?
- What are the implications of these results for policy making in the fields of climate change mitigation and resource use?

Results

This section presents the results of the analyses with a focus on the BRICSA countries (see Tab. 1). The analyses examine the following areas: global flows of embodied carbon emissions and materials; carbon and materials trade balances; and per capita production and consumption.

Carbon Emissions

Table 1 outlines the carbon trade balances, i.e. the differences between carbon emissions embodied in imports and exports, for the years 1995, 2000, and 2005 for the individual BRICSA countries, the group as a whole, the OECD, and the rest of the world (RoW) in million tons (Mt) CO_2.

Table 1: Carbon emissions embodied in trade and consumption for 1995, 2000 and 2005
Abbreviations: CEE = Carbon emissions embodied in exports; CEI = Carbon emissions embodied in imports; RoW = rest of world

in million tons CO_2	China	India	Argen-tina	Brazil	South Africa	Russia	BRICSA	OECD	RoW
1995									
Production	2,993	789	117	244	289	1,595	6,026	10,863	6,031
Consumption	2,285	703	136	284	242	1,072	4,721	12,490	5,709
CEE	829	125	12	24	64	615	1,589	368	1,123
CEI	121	39	31	64	17	92	283	1,995	801
Net-trade	708	86	-19	-40	47	523	1,305	-1,627	322
2000									
Production	3,052	984	134	315	310	1,526	6,322	11,687	6,375
Consumption	2,305	890	148	365	243	886	4,837	13,856	5,691
CEE	915	174	18	40	96	736	1,862	446	1,390
CEI	168	80	31	90	29	96	378	2,615	706
Net-trade	747	94	-13	-50	67	640	1,484	-2,169	684
2005									
Production	5,090	1,171	141	341	341	1,546	8,632	12,138	7,315
Consumption	3,665	1,037	122	319	292	1,041	6,478	15,241	6,366
CEE	1,734	245	35	84	91	675	2,702	519	1,679
CEI	309	111	16	61	43	171	548	3,622	730
Net-trade	1,425	134	19	23	48	504	2,154	-3,103	949

The results highlight that both China and Russia were by far the largest carbon producers and consumers within the BRICSA countries in 1995. Examined on a per-capita basis, Russia and South Africa had the highest per-capita consumption and production, as outlined in Figure 2.

In contrast to the other BRICSA countries which all recorded increases in most of the categories, Russia experienced a decline in CO_2 production and consumption and net exports in the timeframe from a high in 1995 until 2005. This was most likely due to continuing economic impacts from the collapse of the Soviet Union in 1991. Despite this, Russia still ranked a far second behind China in all categories by 2005. In contrast, China's consumption and production increased by 60% and by 70% respectively in the same timeframe.

Overall, total CO_2 production in the BRICSA countries increased by 49% and consumption by 43% in the same timeframe. However, OECD countries' CO_2 consumption increased far more than CO_2 production (24%, or 2,689 Mt CO_2 versus 13%, or 1,213 Mt CO_2).

The results indicate that much of the increase in carbon consumption in the OECD countries was absorbed by the BRICSA countries, where carbon imports doubled between 1995 and 2005. In contrast, there were minimal changes to CO_2 embodied in exports from OECD countries to the BRICSA countries in the same decade (see Figure 1). CO_2 emissions embodied in exports from the BRICSA countries to the rest of the world (RoW) exceeded CO_2 emissions embodied in imports from these countries. Embodied CO_2 emissions are displayed in Figure 1.

Figure 1: CO_2 embodied in imports and exports of the BRICSA countries, 1995 to 2005, Source: Wiebe et al. 2012b

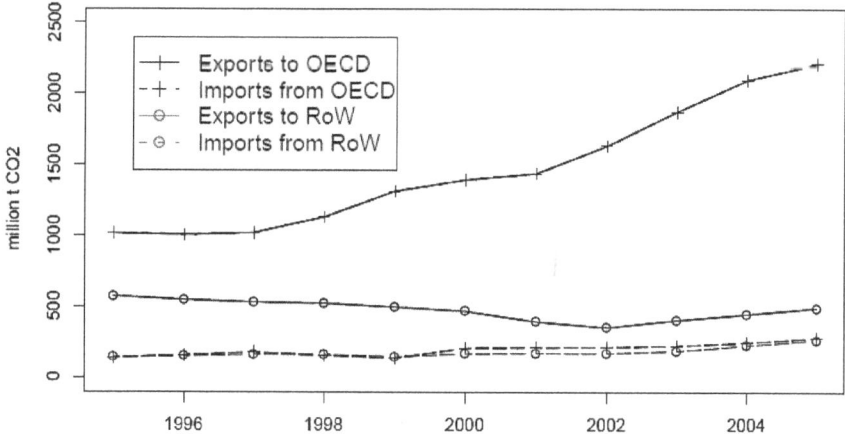

Figure 2 outlines the per capita carbon production and carbon consumption for the three groups: RoW, OECD and BRICSA, and for the individual BRICSA countries. The results highlight that OECD carbon production per capita (about 9 t CO_2) was significantly lower than its per capita carbon consumption which increased from around 10.5 t CO_2 in 1995 to 11.5 t CO_2 in 2005). In contrast, carbon production per capita in the rest of the world was slightly higher than carbon consumption (around 2 t CO_2 compared to 1.8 t CO_2). Overall, in this timeframe, per capita carbon production was found to be higher than carbon consumption in all BRICSA countries, with Russia emitting by far the largest amount of CO_2 per capita.

Figure 2: Carbon emissions per capita from a production and a consumption perspective, Source: Wiebe et al. 2012b

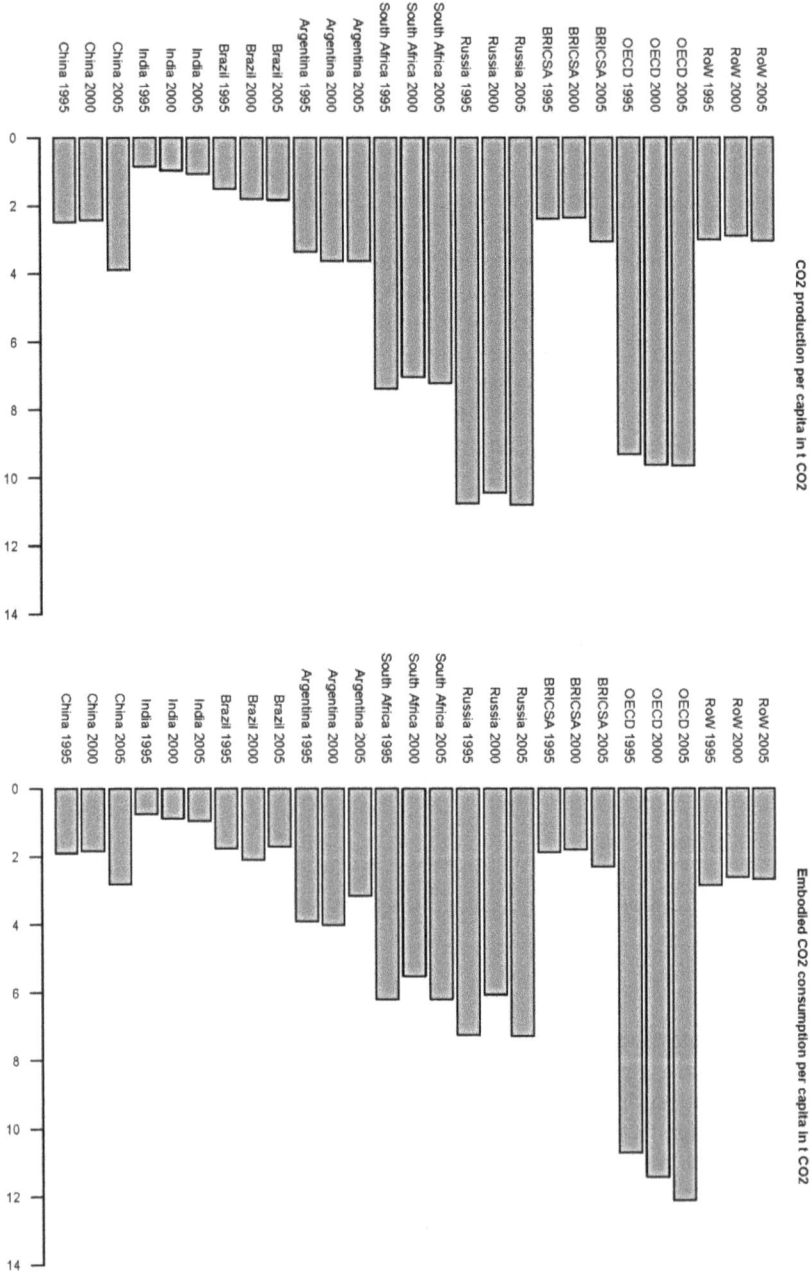

Material Extraction

As well as examining carbon emissions embodied in international trade, GRAM also allows for the calculation of the materials embodied in 4 aggregated categories: biomass; fossil fuels; metals and industrial minerals; and construction materials. Again, the results highlight that exports from the BRICSA countries and the RoW contained more embodied materials than imports, meaning that they are net exporters of embodied materials while the OECD countries are net importers.

While material extraction in the OECD countries only increased by 7% between 1995 and 2005, both BRICSA and the RoW increased material extraction by about 35%. The highest absolute rise in material extraction in this timeframe was observed in China where extraction increased to a total of 3 billion t. This increase by itself almost covers the increase in net imports of embodied material into the OECD countries, however, at the same time domestic consumption in China increased by almost 2.5 billion tons, so net exports only rose by 413 Mt.

Table 2: Materials embodied in trade and consumption for 1995, 2000 and 2005
Abbreviations: MEE = Materials embodied in exports; MEI = Materials embodied in imports; RoW = rest of world

in million tons	China	India	Argen-tina	Brazil	South Africa	Russia	BRICSA	OECD	RoW
1995									
Domestic extraction	5,155	2,561	732	2,310	618	2,009	13,386	20,746	12,293
Consumption	4,234	2,298	689	2,263	557	1,557	11,597	25,173	9,655
MEE	1,181	374	108	224	101	772	2,554	580	3,448
MEI	259	110	65	177	40	320	765	5,008	809
Net-trade	922	264	43	47	62	452	1,789	-4,427	2,638
2000									
Domestic extraction	5,709	2,811	833	2,492	669	1,971	14,485	22,167	14,066
Consumption	4,822	2,616	766	2,378	566	1,068	12,217	27,966	10,536
MEE	1,279	405	143	303	168	1,108	3,126	643	4,380
MEI	391	210	76	189	66	206	857	6,442	850
Net-trade	888	195	67	114	102	902	2,269	-5,799	3,530
2005									
Domestic extraction	7,995	3,129	895	3,006	711	2,341	18,077	22,824	16,535
Consumption	6,660	2,951	637	2,575	656	1,546	15,025	30,327	12,085
MEE	2,028	502	305	604	168	1,218	4,311	737	5,479
MEI	693	324	46	173	113	424	1,258	8,241	1,028
Net-trade	1,335	178	258	431	56	795	3,053	-7,503	4,451

The increase in materials embodied in trade between BRICSA and OECD countries (see Figure 3) was similar to that of the increase in carbon exports outlined in Figure 1. During this time, embodied biomass and fossil fuel imports from the OECD countries remained more or less constant, while exports to OECD countries rose significantly. The same is true for exports of metals, industrial and construction minerals from BRICSA to OECD countries. In contrast, import flows of these materials increased only marginally.

Materials embodied in imports from OECD countries were between 50 and 100 Mt per year per material category from 1995 to 2005. However, the extent of materials embodied in exports to OECD countries differed between material categories, with MEEs of biomass and fossil fuels increasing from around 800 Mt to around 1,200 Mt and 1,500 Mt, respectively and MEEs of metals and industrial minerals increasing from around 300 to 580 Mt with construction minerals increasing from around 200 to around 400 Mt in the same timeframe. Total MEEs to OECD countries increased by 85% from 2 billion t in 1995 to 3.7 billion t in 2005, while MEIs only grew by 50% from 230 to 340 Mt.

Figure 3: Biomass, fossil fuels, metals and industrial minerals, and construction minerals embodied in imports and exports of the BRICSA countries, 1995 to 2005, Source: Wiebe et al. 2012b

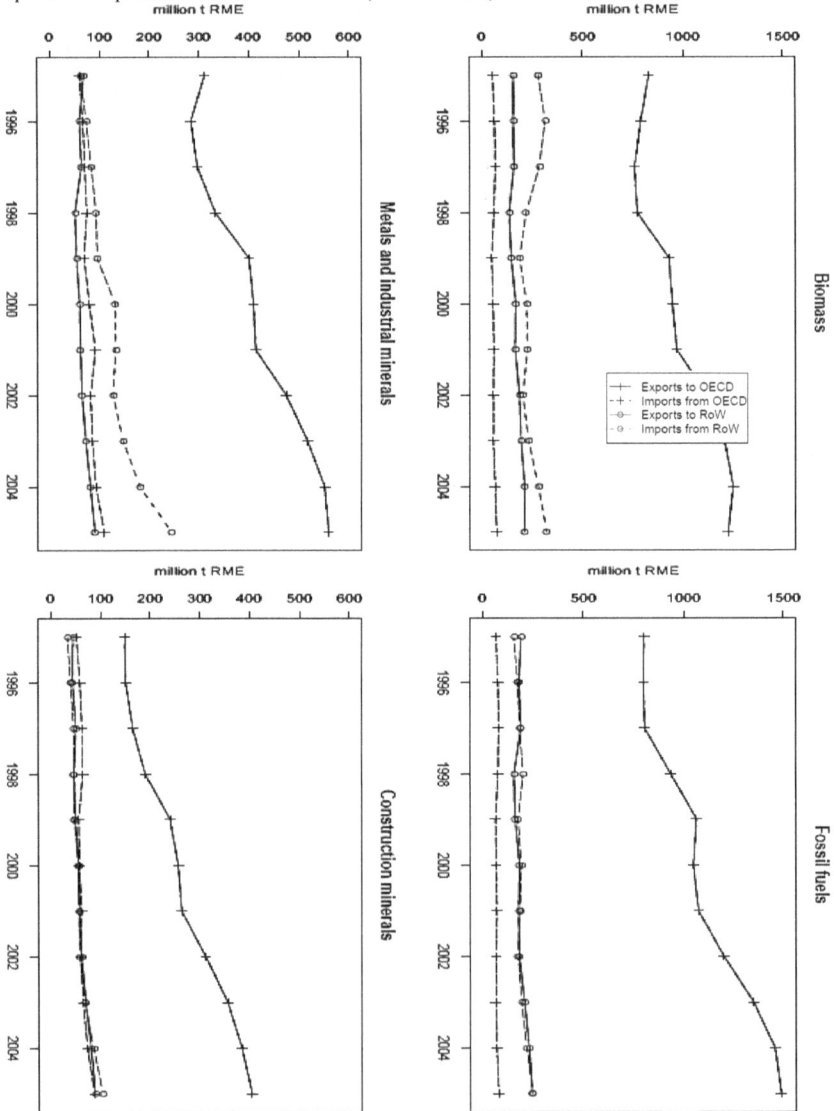

Figure 4: Material extraction and consumption per capita, Source: Wiebe et al. 2012b

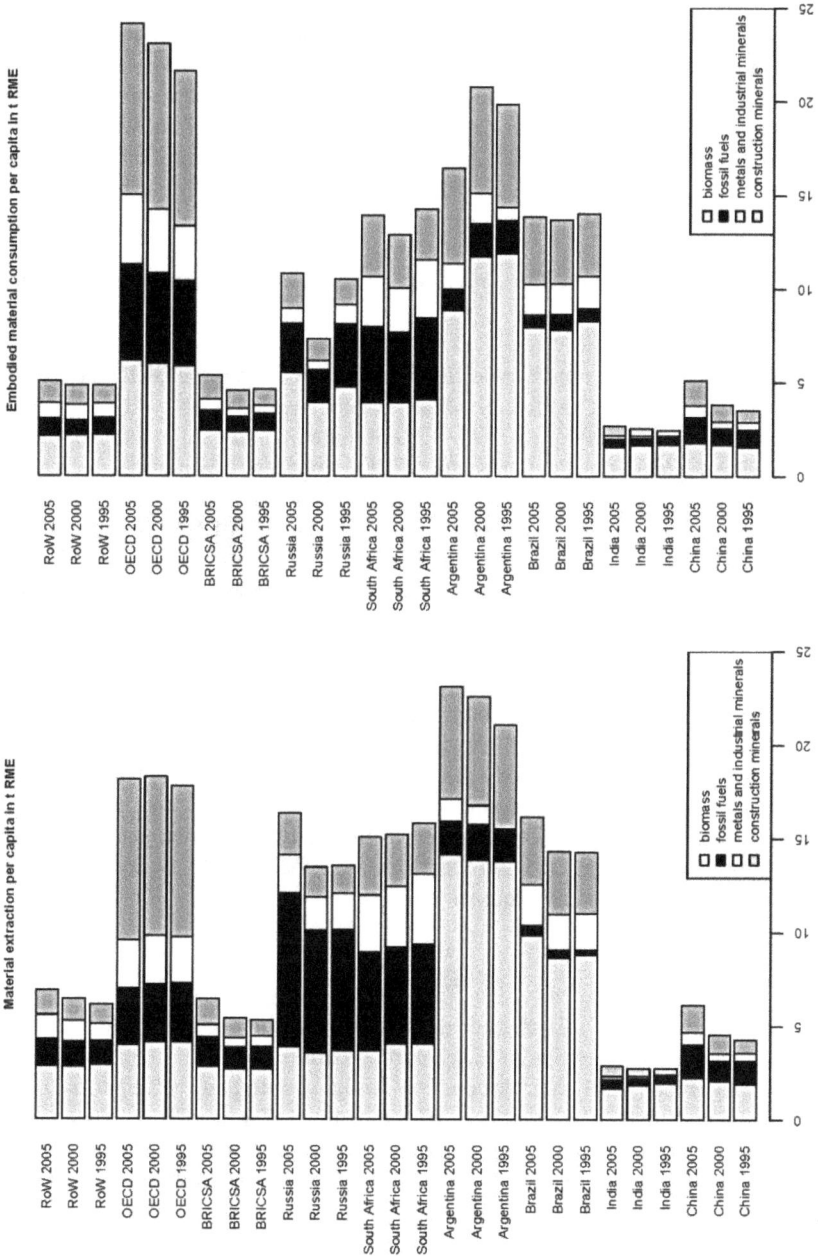

While carbon exports from BRICSA countries to the RoW were higher than carbon imports, this relation does not hold for embodied materials where imports of biomass, metals and industrial minerals to BRICSA countries exceeded exports from these countries. There was a particularly significant increase in imports of metals and industrial minerals from the RoW to BRICSA countries, while exports from BRICSA countries only increased slightly, if at all. Imports of embodied materials to and exports from the RoW were almost equal for fossil fuels and construction minerals.

Figure 4 shows the per-capita composition of material extraction and material consumption where again, per-capita consumption of materials was lower than material extraction in the BRICSA countries, whereas consumption in the OECD countries was significantly higher than extraction, particularly for biomass, fossil fuels, metals and industrial minerals.

OECD's influence on carbon emissions and material extraction in the BRICSA countries

The results in the previous two sections demonstrated that the BRICSA countries were net exporters of both embodied carbon emissions as well as embodied materials from 1995-2005. In 1995, 80% of material and carbon exports from the BRICSA countries were exported to OECD countries; this number increased to 85% for both material and carbon embodiments in 2005. There was a significant increase in the share of carbon emitted in the BRICSA countries during production of goods for exports to the OECD countries from 20% in 1995 to 28% in 2005, meaning that the OECD countries were responsible for more than one quarter (2,210 Mt CO_2) of emissions produced within the BRICSA countries (8,632 Mt CO_2).

Thus Figures 2 and 4 demonstrate that refocusing responsibility for emissions from the producing countries to the consuming countries dramatically changes the way in which the responsibility for environmental pressures can be apportioned. Therefore, applying the MRIO-derived indicators considerably shifts responsibility to OECD countries, while improving the environmental performance for the BRICSA countries, as the results show that per capita carbon emissions and material consumption increased for the OECD and at the same time decreased for BRICSA countries.

Discussion

The preceding part of this chapter discussed the results of the application of the GRAM model on consumption-based carbon emissions and material consumption, contrasting them with their production-based counterparts – territorial carbon emissions and material extraction – in the context of international trade with the BRICSA emerging economies. The results demonstrate that there would be significantly different outcomes in regards to responsibility for environmental pressures if one were to apply a consumption-based accounting system as opposed to a territorial-based accounting system.

The analyses indicate that OECD countries came close to stabilising their levels of domestic emissions and material extraction between 1995 and 2005. However, this coincided with an "outsourcing" of environmentally intensive stages of production and industries in general, to developing and emerging economies with arguably less environmental and human rights protections.

Furthermore, the data also indicate that an increase of CO_2 emissions and material extraction in BRICSA countries correlated at least in part to an increase in consumption in the OECD countries.

An increasing amount of studies are acknowledging that increased resource use and extraction are exacerbating the already strained capacity of the planet to provide resources and absorb waste. Hence, the results of this study, supported by other rigorous analyses in this field, should be used to inform national, regional and international discussions to address the issue of resource over-consumption as well as identify measures to drastically mitigate this situation.

What is required is global responsibility and global action on resource consumption and resource extraction. However, to-date global agreements on parallel areas such as climate change and global trade have reached a stalemate. Therefore, it is hoped that progressive countries and corporations in both the developed and developing world have the courage and foresight to act unilaterally, or cooperate with like-minded countries and entities to integrate the consumption perspective in policy design and business plans to prevent increases or shifts in emissions and material extraction.

Both multilateral and bilateral trade agreements remain a vital basis and opportunity for integrating responsibility of consuming and producing countries and addressing environmental and human rights concerns in the producer country. Furthermore, as more data for other resource categories such as land, water, and biodiversity become available, future model analyses could also focus on these resources. This will deliver further improvement in understanding environmental dependencies that future policy approaches should better take into account.

Conclusions

The consumption perspective reveals increasing environmental pressures, particularly in emerging economies stemming from consumption activities within OECD countries. Emerging and developing countries demand that the OECD countries take action first in reducing global environmental pressures and are only willing to implement measures themselves if the OECD countries provide clear commitments. The current difficult negotiations in the context of global climate treaties illustrate this situation. Respecting that consumers are at least partly responsible for the environmental pressures caused as a consequence of consumption activities, an approach of shared responsibility between producer and consumer countries could help unlock the current difficulties in global negotiations. MRIO models and indicators derived therewith form an essential knowledge base for future policy discussions and environmental negotiations.

Further reading

Bruckner, M., Giljum, S., Lutz, C., & Wiebe, K. (2012). Materials embodied in international trade – Global material extraction and consumption between 1995 and 2005. *Global Environmental Change, 22*, 568-576.

Giljum, S., Lutz, C., Jungnitz, A., Bruckner, M., & Hinterberger, F. (2011). European Resource Use and Resource Productivity in a Global Context. In P. Ekins, S. Speck (Eds.), *Environmental Tax Reform (ETR) - A Policy for Green Growth* (pp. 27-45). New York: Oxford University Press.

Wiebe, K., Lutz, C., Bruckner, M., & Giljum, S. (2012a). Calculating energy-related CO2 emissions embodied in international trade using a global input-output model. *Economic Systems Research, 24* (2), 113-139.

Wiebe, K., Lutz, C., Bruckner, M., Giljum, S., & Poliz, C. (2012b). Carbon and Materials Embodied in the International Trade of Emerging Economies: A Multi-regional Input-Output Assessment of Trends between 1995 and 2005. *Journal of Industrial Ecology, 16* (4), 636–646.

Resources

SERI (2012). Global Material Flows Database. 2012 edition. Available at www.materialflows.net.

Acknowledgements

The results presented in this chapter were sourced from two projects funded by the Austrian Climate and Energy Fund and the Anglo-German Foundation.

Part V: The Role of MRIO in Global Governance

Dingo dreaming
Oil on canvas, 51cm x 65cm
Dagmar Hoffman

Chapter 21: Policy Discussions Using Inter-Country Input-Output (ICIO) Systems[1]

Norihiko Yamano and Colin Webb

Evolution of Global Production Networks

Globalization and international economic integration are not new phenomena. However, in the last two decades the flow of goods between countries has increased dramatically (Figure 1), and many countries have become increasingly interdependent.

Figure 1: World export-to-GDP ratio

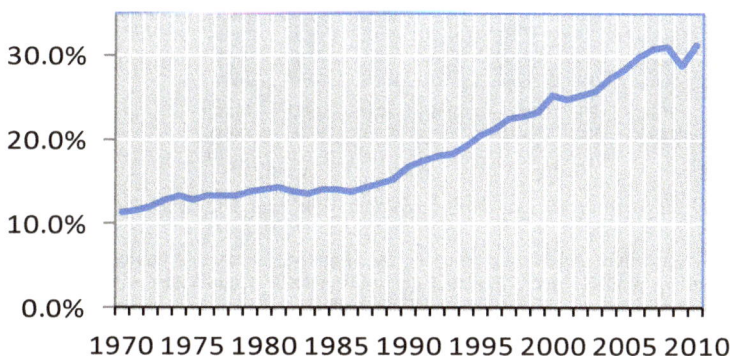

Source: UN National Accounts

Underlying this general trend are trading patterns that have changed significantly - with many countries experiencing notable shifts in the structure of their imports and exports. Figure 2 highlights the evolution of trade structures of major exporting countries. Since the mid-1990s, China's export share of household consumption goods has declined while its export share of capital goods and ICT products, such as personal computers and mobile phones, has increased. However, its share of exports of intermediate goods has remained steady at around 40%.

---

[1] This represents the views of authors and not necessarily those of the OECD

Meanwhile, countries such as Germany, Japan and United States have maintained relatively high export shares of industrial intermediate goods (50-60%) while their export shares of household goods have remained lower than those of China. Although export shares by end-use categories have remained relatively stable in these OECD countries, there have been structural changes concerning the product groups. The US export shares of machinery and equipment have increased both for intermediate and capital goods, while the shares of these products have decreased in Germany.

Figure 2: Export shares by end-use for four major exporting countries, 1995 and 2011

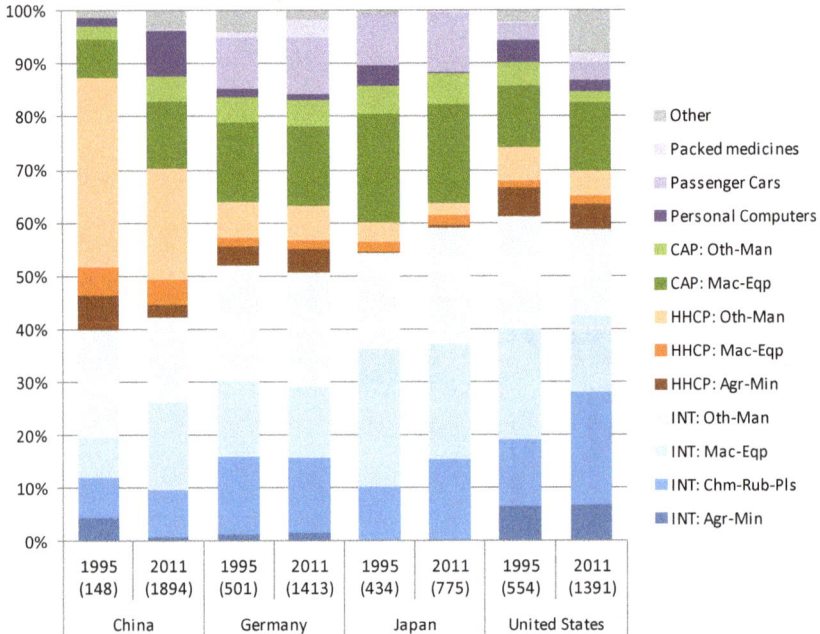

Source: OECD Bilateral Trade Database by Industry and End-Use categories (BTDIxE), 2012

Note: Agr-Min: Agriculture and Mining products; Mac-Eqp: Machinery and Equipment; Oth Man: Other Manufacturing products; INT: Intermediate inputs; HHCP: Household Consumption; CAP: Capital goods. Certain product groups with mixed end-uses are shown separately. Figures in parentheses are total exports in USD Billion

Changes in trade structure are highly correlated with the evolution of global supply chains i.e. international procurement activities. In other words, imported manufactured parts and components are increasingly used in production processes of each country. The share of imported goods and services in the total intermediate inputs of OECD countries increased from 12.2% in the mid-1990s to 16.2% in the mid-2000s[2].

---

[2] OECD Input-Output Database (www.oecd.org/sti/inputoutput)

However, the share of intermediate import contents of final expenditure can vary widely across industries due to the variety and availability of natural resources, industrial supplies and supporting business services (Table 1).

Table 1: Intermediate import contents share of unit final expenditure, Germany

Industry	1995	2005	Industry	1995	2005
Agriculture, hunting, forestry and fishing	12%	17%	Other transport equipment	35%	33%
Mining and quarrying	11%	23%	Manufacturing n.e.c; recycling	20%	27%
Food products, beverages and tobacco	19%	23%	Electricity, gas and water supply	8%	19%
Textiles, textile products, leather and footwear	32%	37%	Construction	11%	16%
Wood and products of wood and cork	15%	21%	Wholesale and retail trade; repairs	6%	8%
Pulp, paper, paper products, printing and publishing	18%	20%	Hotels and restaurants	12%	13%
Coke, refined petroleum products and nuclear fuel	60%	76%	Transport and storage	12%	19%
Chemicals and chemical products	22%	31%	Post and telecommunications	6%	14%
Rubber and plastics products	23%	28%	Finance and insurance	7%	10%
Other non-metallic mineral products	12%	19%	Real estate activities	2%	4%
Basic metals	32%	42%	Renting of machinery and equipment	1%	0%
Fabricated metal products	18%	23%	Computer and related activities	5%	8%
Machinery and equipment n.e.c	19%	25%	Research and development	8%	11%
Office, accounting and computing machinery	31%	39%	Other Business Activities	4%	6%
Electrical machinery and apparatus n.e.c	17%	25%	Public administration; social security	5%	7%
Radio, television and communication equipment	28%	36%	Education	3%	3%
Medical, precision and optical instruments	14%	20%	Health and social work	5%	7%
Motor vehicles, trailers and semi-trailers	25%	34%	Other community, social and personal services	6%	7%

Source: OECD Input-Output Database

Conventional national input-output models treat export and import activities as exogenous factors in a state's economy. Economic modelling using Inter-Country Input-Output (ICIO) systems can provide insights into international spillover and feedback effects. As countries have become increasingly dependent on external markets, the ICIO type input-output model has regained the attention of policy makers. In particular, environmental and trade policy analysts are supplementing their conventional approaches with information concerning international spillover effects drawn from ICIO systems.

Relevant Policy Areas

With its specific information on international transactions between target countries, a good ICIO database can be useful for a range of policy analyses. In particular, indicators can be developed to inform discussions in the areas of trade policy, industrial policy, environmental policy and risk management, as well as contribute to international trade negotiations

Trade Policies

Indicators based on an ICIO system can provide both conventional and alternative viewpoints of our understanding of bilateral trade relationships, the depth of interdependency due to international supply networks as well as new measures of comparative advantage and productivity. For example, conventional bilateral trade balances based on the total reported value of exports and imports can be misleading due to the significant international flows of manufactured intermediates. As an alternative to these familiar "gross measures", an ICIO system can be used to develop measures of trade in value added terms i.e. reveal how much domestic value added is present in reported exports to a particular partner country or, present in the latter's consumption of final goods and services.

There are various areas where measuring trade in value-added may bring new perspectives. For this reason, and to meet increasing international demand, the OECD, in conjunction with WTO, has embarked on a long-term project to develop and maintain a range of trade in value added indicators to better inform policy makers[3]. Trade analysts are particularly interested in using the new information to address the following issues:

1. Global imbalances: With respect to a country's overall trade surplus or deficit with the rest of the world, measures based on conventional 'gross' trade flows and value-added measures are consistent, but on a bilateral basis, conventional statistics on flows of goods and services, can present a misleading picture of who ultimately benefits from the trade, and exaggerate the importance of producing countries at the end of value chains. Value-added measures of bilateral trade better reflect who benefits, both in monetary terms but also, by extension, employment terms.

2. Market access and Trade disputes: Conventional measures may create a risk of protectionist responses that target those countries at the end of global value chains, on the basis of an inaccurate perception of the origin of trade imbalances. Indeed 'beggar thy neighbour' strategies can turn out to be 'beggar thyself' miscalculations.

3. Managing macro-economic shocks: The 2008-2009 financial crisis was characterised by a synchronised trade collapse in all economies, as the effects of a drop in demand fed through to countries located upstream in the global value chain. A better understanding of value-added trade flows would provide tools for policymakers to identify the transmission of macro-economic shocks and adopt the right policy responses.

4. Trade, growth and employment: While there are concerns that imports threaten domestic jobs, the reality is that jobs are increasingly created as part of global value chains. Trade flows in value-added terms indicate where jobs are created and highlight the benefits of trade for all economies involved in the value chain. Interdependencies within global value chains are key to explaining the competitiveness of countries and the productivity gains that capitalise on these dependencies.

---

[3] See www.oecd.org/trade/valueadded and
www.wto.org/english/res_e/statis_e/miwi_e/miwi_e.htm

Environmental Issues

Changes in consumption and production locations have significantly altered the global patterns of consumption-based ecological impacts and production-based ecological impacts. For example, efforts to mitigate greenhouse gas (GHG) emissions, such as the Kyoto Protocol, will be less effective in reducing global emissions of GHG if countries with emission commitments relocate their carbon-intensive production activities to countries without such commitments, particularly if production in the latter countries is GHG-intensive. An ICIO combined with IEA energy statistics (e.g. fuel-combustion-based CO_2 and international electricity transfer), and other industry statistics, can be used to estimate the effects of international transfers of CO_2 emissions.

OECD has long been involved in policy analysis concerning the environment and sustainable development - most recently under the umbrella of the Green Growth project[4]. One of the key indicators identified concerned embodied CO_2 in international trade to highlight CO_2 consumption patterns as an alternative to the widely used production-based measures of CO_2 emissions. Growth of trade-adjusted consumption-based CO_2 emissions has not fallen in line with domestic CO_2 emissions in OECD countries, partly reflecting the increased global sourcing of emissions-intensive activity from non-OECD countries. While less than half of the global increase in CO_2 emissions during the second half of the 1990s came directly from OECD economies, two-thirds of the global increase was attributable to OECD consumption. By the mid-2000s, the regional contributions to increasing emissions had shifted so that about 12% of the global increase in CO_2 emissions between 2000 and 2005 came directly from OECD economies with a quarter attributable to OECD consumption.

Moreover, ICIOs are useful for analysing local pollution and ecological footprints. For example, analysis of freshwater extraction, soil degradation and biodiversity impacts are some of the subjects that have turned to ICIO models for insights (e.g. EXIOPOL, Lenzen M, Kanemoto K, Moran D, and Geschke A (2012) Mapping the structure of the world economy, Environmental Science and Technology 46(15) pp 8374–8381).

Risk Management

Recent unexpected and devastating events such as the March 2011 earthquake in Japan and the autumn 2011 flood in Thailand raised some understandable concerns over global supply chains. The rise of global supply chains has increased the sensitivity of national economies to natural disasters or other shocks in other parts of the world. ICIO models can contribute to a better understanding of direct and indirect vulnerability to, and consequences of, unexpected events, to inform countries about possible pre-emptive actions to minimise impacts.

---

[4] www.oecd.org/greengrowth

What Does an Inter-Country Input-Output (ICIO) Model Represent?

In the discipline of regional science, Multi-Regional Input-Output (MRIO) frameworks have been discussed at length and developed for many policy applications – particularly in larger economies such as Japan and USA Sub-national regions are relatively open and depend on each other more than international trade partners[5]. However, while international transactions are influenced by some of the same factors as sub-national inter-regional transactions such as transport costs, they can face certain impediments such as tariffs and non-tariff barriers outside of a Regional Trade Agreement (RTA) framework.

Building on the techniques developed for and applied to MRIOs, various research teams have undertaken efforts to develop Inter-Country Input-Output (ICIO) models. Current ICIO projects include:

IDE-JETRO: 9 Asian economies and United States, 1985-2005

http://www.ide.go.jp/English/Publish/Books/Sds/material.html

EXIOPOL: 43 countries, 2000

http://www.feem-project.net/exiopol/

OECD Inter-Country Input-Output Database: 34 OECD countries and 23 other economies, 1995-2009

http://www.oecd.org/sti/inputoutput/

University of Sydney EORA MRIO: 187 countries, 1990-2010

http://www.worldmrio.com

World Input-Output Database: 27 EU countries and 14 other economies, 1995-2009

http://www.wiod.org

The underlying data sources for an Inter-Country Input-Output (ICIO) system are national Input-Output (IO) or Supply-Use (SU) tables, National Accounts series and bilateral trade coefficients that, ideally, have been harmonised – for example, to cover a common industry list. The resulting ICIO then contains comprehensive information concerning industrial activities such as international trade, consumption and investment by activity and sales and procurement information.

An ICIO based on an adequate number of countries and industrial detail can therefore be useful as a data source for identifying international and sectoral transactions. It can even provide alternative (improved) gross trade figures of goods and services than those publicly available from official statistics.

---

[5] Anderson JE and van Wincoop E (2044) Journal of Economic Literature Vol. XLII pp. 691-751.

Methodological and Statistical Limitations

As discussed above, the ICIOs provide useful numerical indicators for evidence based policy making in various areas. However constructing such an ICIO table with maximum global coverage requires extremely data intensive preparation procedures. The key issues are summarized as follows.

a) *Collection and estimation of Input-Output tables in harmonized format.* While national tables are expected to comply with a standard (industry/product) format within Europe, most other countries produce SUTs / IOTs in their own established formats with varying industry/product detail. Also, the definition ('valuation') of output can vary across countries e.g. treatment of taxes and subsidies.

b) *Development of bilateral trade in goods and services consistent with National Accounts framework of each country.* Reported bilateral merchandise exports and imports statistics compiled by Customs offices are not, by definition, the same as estimates of exports and imports of goods in Balance of Payments (BoP) and National Accounts (SNA) statistics and hence national I-O tables. National statisticians make numerous adjustments to transform merchandise trade in goods statistics to SNA/ BoP concepts. For example, adjustments are made for the "cost, insurance, freight" (c.i.f.) element of reported merchandise imports of goods. However, for many countries such adjustments are carried out only for total goods, or at least at very aggregate levels. Thus assumptions are required to link bilateral trade by industry or product groups to I-O tables. Also, adjustments may be required to account for the presence of re-exports or re-imports in official bilateral trade statistics as well as to deal with recorded flows of used (second-hand) and recycled products.

c) *Estimation methodology of international trade and insurance margins and tariffs.* International trade and insurance margins need to be explicitly estimated to link the monetary transactions between countries. The actual international transaction costs are determined by various factors such as distances between countries, port efficiency, type of commodity and fuel surcharge prices.

d) *Incorporating and understanding firm heterogeneity.* Statistics from customs databases in China and Mexico suggest that the majority of exports from certain sectors are driven by foreign capital with many firms serving export markets only. Such (processing) firms may have very different characteristics from other firms in the same sector serving their domestic market. Accounting for such heterogeneity within certain sectors is a particular challenge when building an ICIO model.

OECD has a long history in the development of harmonised national input-output tables and bilateral trade in goods by industry databases for analytical use. These unique data sources together with national statistics such as annual National

Accounts and bilateral trade in services have been used for a range of country comparative analyses - such as those alluded to above.

OECD's new and evolving ICIO database is based on methodologies previously established for inter-regional analyses (e.g. Chenery-Moses; Isard). To populate the OECD ICIO, linking national I-O tables with international trade coefficients, the following statistics are collected and adjusted.

Official statistics submitted by OECD members and major non-OECD countries	Equivalent OECD harmonised analytical databases	Adjustments for ICIO development
Annual National Accounts (SNA)	Structural Analysis Database (STAN)	SNA benchmarked I-O and trade in goods and services
Input-Output / Supply use tables	OECD Input-Output database	Re-export adjustments for trade in National accounts
International merchandise trade (UN Comtrade / OECD ITCS)	Bilateral Trade Database by Industry and by Enduse (BTDIxE)	Imputed trade flows for goods and services
Balance of Payments	Bilateral Trade in Services	Estimation of updated tables for reference years using close years information

Summary

New indicators based on ICIO systems can support evidence-based policy making processes particularly for areas related to productivity and competitiveness. New information on comparative advantages can help countries better understand global production networks and determine optimal strategies for taking advantage of resource efficiencies within global production chains.

The framework of an ICIO system still follows the principles of the System of National Accounts (SNA) so the model is basically closed in monetary transactions between sectors of a specific reference year. Physical transactions, capital formation over years, income transfers such as payments for royalties, dividends and labour income dividends are not yet in the scope of ordinary ICIO models. This additional type of statistical data information integrated within an ICIO system would provide deeper understanding of the evolution of global production networks.

Chapter 22: The Use of MRIO for multilateral trade policy

Christophe Degain, Hubert Escaith, and Andreas Maurer[1]

Introduction

Further to technological progress and market opening, new business models emerging in the last decades have increased the international fragmentation of production. International transactions have become a complex nexus of investment-trade-services in global production networks (GPNs) with an increased macroeconomic interdependence between economies. The notion of trade in tasks has emerged, with the distribution of activities along the stages of international production chains. These changes highlight the need for new measures of international trade to take into account the reality of the globalized economy. The *trade in value added* measure is one of them. It relies on Multi-Region Input-Output (MRIO) tables (see Chapter 1). This measure helps us to understand the economic and social dimensions of trade and brings new perspectives to the design of trade policies and to the analysis of their impact.

---

[1] This Chapter has been written in the personal capacity of the authors and does not represent the views of the WTO members or Secretariat

MRIO to Better Understand Trade in the 21st Century

From Trade in Goods to Trade in Tasks

Today, most manufactured products for mass consumption are produced within international supply chains with companies specializing in specific tasks or stages of the production process. Thus, more and more products are composed of parts and accessories (intermediate goods) that come from various geographical origins. The label on the back of such products should be 'Made in the World' rather than 'Made in country A'[2]. Grossman and Rossi-Hansberg (2006) describe this as *trade in tasks*.

In 2010, intermediate goods exports represented around 54% of world non-fuel merchandise exports. The value of these inputs embedded in goods is in fact counted each time they cross borders, thus, trade in intermediates results in multiple counting in gross statistics. This is why an increased trade elasticity – world trade growing more rapidly than world GDP – has been observed since the late 1980s. Though, it seems that these increases in trade elasticity are only temporary with a progressive return to the long-term average[3].

Trade in intermediate goods is however only the visible part of the iceberg of GPNs. Other features include the increased use of services and the important growth of intra-firm transactions and inter-action, including financial services and knowledge sharing.

As the complexity of products increases, especially in the electronics sector, GPNs have broadened to include developing countries, with an important part of this trade taking place in export processing zones (EPZs). These are geographical areas benefiting from special administrative and fiscal status aimed at promoting investment and trade. Developing economies often use these EPZs to attract production facilities which process imported inputs that are afterwards exported. Some 20% of their merchandise exports come out of EPZs.

Services are the lubricant of GPNs appearing at different stages of the production process as shown by the 'Smiley' Chart in Figure 1.

For example, infrastructure services such as transport, communication or finance are needed to create the conditions for ensuring the smooth operation of supply chain production. Other services ('intermediate services'), such as design or computer services, etc. are part of the production process at every stage of the chain and 'embodied' in the intermediaries as these are passed on to downstream processors. Further services are required for marketing, sale and, possibly, repair of the merchandise. These may be called embedded services.

---

[2] See WTO's "Made in the World" Initiative (MIWI) at http://www.wto.org/english/res_e/statis_e/miwi_e/miwi_e.htm, last accessed at 25 April 2012. See also WTO/OECD's initiative to develop trade in value added statistics at http://www.wto.org/english/news_e/news12_e/miwi_15mar12_e.htm, last accessed at 21 May 2012.

[3] Escaith, H., Lindenberg, N., & Miroudot, S. (2010). *International Supply Chains and Trade Elasticity in Times of Global Crisis*, Geneva: WTO, Staff Working Paper N° ERSD-2010-08.

Figure 1: An illustration of the role of services in the manufacturing process

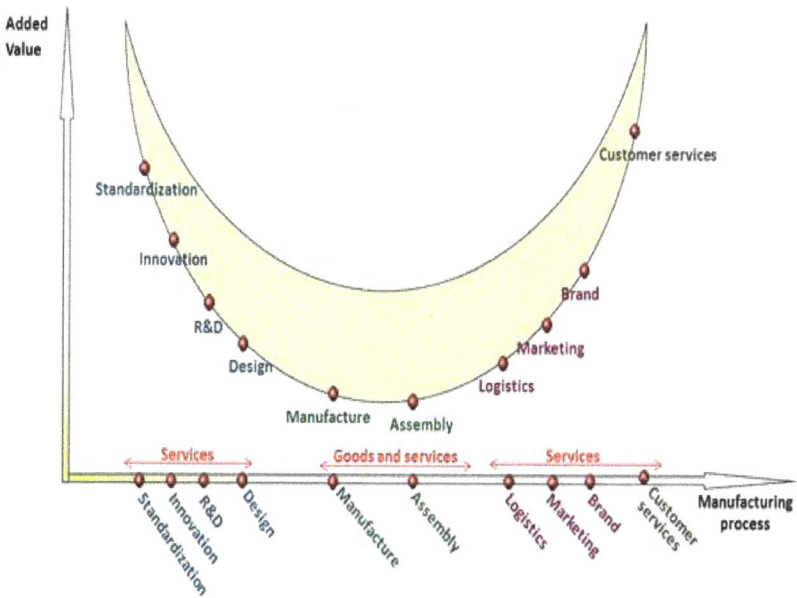

Source: WTO, based on Shih S., Business Week (May 16, 2005).

Many services are actually indirectly traded through goods as the distinction between embodied and embedded is not always clear-cut, that is, the differentiation between goods and services is blurred. For example, the flow of intermediate goods between various supply chain actors implicitly reflects the exchange of business and manufacturing services. The contribution of these service suppliers, often small and medium enterprises (SMEs), is ignored by traditional trade in merchandise statistics. However, taking into consideration the sectoral origins of inputs included in exports reveals that (i) services constitute a large component of 'manufactured' exports, especially in industrialised countries, and (ii) the participation of SMEs is much higher than what is usually perceived. In the US, the SMEs' participation in world trade goes up from 30% to more than 40% when trade flows are estimated in value added terms[4].

Also, vertical integration of multinational corporations has increased. Intra-firm trade, which represents the exchanges taking place between parent companies and their foreign affiliates, has grown markedly. Anecdotal knowledge sets the range of intra-firm trade at global level between 30% and 40%. Complex supply chains require long-term business relationships, involving transfer of technology or sharing of research and development. Most, if not all of these transactions, require long-term institutional cooperation between the actors. Thus, improved border norms – the usual domain of trade negotiations – demand in addition the development of harmonized investment and industrial norms. This

---

[4] See A. Jara, WTO DDG Jara urges developing further trade statistics on global value chains, http://www.wto.org/english/news_e/news12_e/ddg_20apr12_e.htm, last accessed at 25 April 2012.

characterizes a 'deep integration' that in turn explains in good part the multiplication of bilateral or regional trade agreements[5].

Towards a Measure of Trade in Value Added

Measuring the evolution of trade related to global production networks is a challenge. Conventional trade statistics obscure the production-related aspects of international trade. For example, the recorded country of origin of imports is often the last country in the production chain. This overstates the value of exports of the last country in the chain since the contributions of the other countries in the chain are ignored. Another bias observed with conventional trade statistics is the multiple counting of transaction values for intermediate goods.

Thus, trade in tasks calls for complementary ways of measuring and analysing trade flows. It requires understanding the global production process and measuring the international exchange of intermediate goods between industries. Calculating value-added in this way is very similar to what national statisticians do to measure the Gross Domestic Product. Taking into account these production structures in estimating trade in value added terms enables us to circumvent the double counting biases faced by conventional trade statistics. We are then better able to evaluate the actual contribution of foreign trade to an economy or to assess more realistically the contribution of services to international trade and to take into account the interconnection of national economies within GPNs.

The statistical tool for this is Multi Regional Input-Output (MRIO) tables that describe cross-border input-output relationships. MRIO tables allow the decomposition of gross exports into their domestic and foreign content. Various value-added components can be derived, such as the domestic value added content included in exports of final or intermediate goods of country A absorbed by importing country B, the domestic value added re-exported by country B to a third country C, the domestic value added re-imported in country A. The domestic and foreign value added components would then add up to the gross value of exports reported in conventional statistics.

Estimating trade in value added has many advantages, the most prominent being the modification of bilateral trade balances with a redistribution of the bilateral trade flows according to the countries of origin of the respective tasks and intermediate inputs contained in a product. In traditional trade statistics, the full value of this good would be attributed to the country of final assembly, which is often the contributor of the low value-added per unit activity, as shown in the 'smiley face', see Figure 1.

Figure 2 compares the geographical origins of the US trade deficit with China in iPhones according to the traditional and the value added measures. Traditional gross statistics attribute the entire value of the iPhone's export to China – as the final assembler in the production chain; hence, the whole deficit is attributed to China. However, the value added estimate highlights Japan, Germany and the Republic of Korea, which are not even considered within gross statistics, as the main sources for the US shortfall in this product. This example

---

[5] See WTO World Trade Report, 2011

also demonstrates that, whatever the statistical method applied, the global trade balance of an economy remains unchanged.

Figure 2: 2009 US trade balance in iPhones (millions US$)

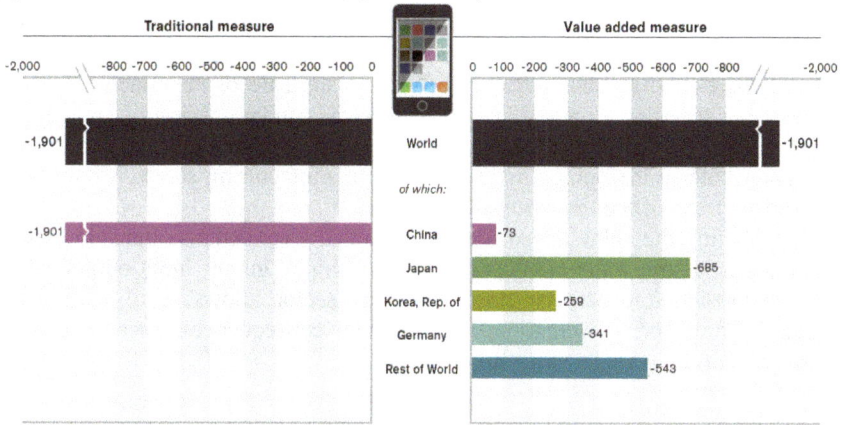

Source: Meng and Miroudot (2011) based on Xing and Detert (2010).

Equally, the value added analysis of trade flows shows that the impact of exchange rate policies in rebalancing bilateral imbalances is limited. For example, if the Chinese value added in US imports stands at 63% (WTO estimate for 2008, including processing trade), an appreciation of the Yuan vis-à-vis the US dollar will only marginally increase the cost of Chinese goods, since the revaluation does not affect the cost of foreign inputs used by Chinese manufacturers. That is, bilateral exchange rate misalignments only partially explain trade imbalances and other macro-economic fundamentals such as saving rates are more important in explaining these imbalances.

Focusing on the domestic value added and its sectoral distribution also impacts on the trade and development relationship. Traditional theories assume that national firms compete through selling finished products based on their overall competitiveness on large markets to benefit from positive returns to scale. In the past, firms from developed countries had a clear advantage over their competitors from developing countries as it was much more difficult for the latter to access large markets. When dealing with international trade generated by GPNs, the structural deficiencies of developing countries are less a handicap for they can perfectly excel in one segment of the production chain, replacing competitors that produce more expensively. Similarly, the size of the domestic market is much less important as there is no need to build cars or airplanes only for the domestic market. Instead developing countries can specialize in activities of a particular segment of the chain that produces a car or a plane for consumption somewhere else in the world. In other words, industrial policies need not be sectoral but rather should focus on operational clusters that involve a series of suppliers from different sectors that produce according to the standards of the production networks

Interestingly, the analysis of international supply chains from a trade and development perspective recaps many of the recommendations of the cluster approach, developed in the 1990s along the lines identified, inter alia, by Michael Porter (1985). For example, identifying the sectoral origin of the domestic content embedded in exports allows for a better identification of the factors that contribute to the competitiveness of nations. In developed countries, in particular, a large share of the value-added embedded in the sale of manufactured goods comes from services. These 'back-office' services can be upstream (research and development) or downstream (communication and logistics). Their quality and cost contribute to the competitiveness of the manufacturing sector, which takes the role of 'front-office' exporter.

Attributing the value-added where it is actually created also helps to expand the analysis of the trade-employment relationship. Linking input-output tables with socio-economic data on employment by sector allows us to analyse the impact of trade on job creation, distributional consequences, or as a general source of gain from trade within an economy.

Global Value Chains and the Design of Trade Policy

The Magnified Impact of Transaction Costs (Tariff, Non-Tariff Measures, Energy, Logistics, Communication, Distribution)

The implications of new measures of trade in value added for understanding competitiveness and for devising pro-trade policies cannot be overestimated. The rise of international supply chains in the 1990s was closely linked to the reduction of international transaction costs, due to progress in communication but also thanks to the improvement in institutional governance, leaner cross-border formalities and the reduction in tariffs over time[6]. Tariffs still matter in some areas, but they have become far less important relative to policy. Today, tariffs are relatively low and most transactions in electronics for example benefit from duty free treatment under the Information Technology Agreement (ITA), a landmark sectoral trade accord signed by many WTO members in December 1996[7]. Increasingly, the source of trade barriers comes from non-tariff measures, such as anti-dumping and domestic regulations, norms and standards[8]. According to the WTO World Trade Report 2011, this trend was behind the rise of regional trade agreements going beyond standard multilateral trade measures to include so-called WTO-X matters that cover investment and other business matters such as competition policies. These WTO-X negotiations were necessary in order to facilitate the deeper integration required by GPNs.

---

[6] Yi, K. M. (2003). *Can vertical specialization explain the growth of world trade?* Journal of Political Economy 111(1): 52-102.

[7] WTO (2012). Fifteen Years of the Information Technology Agreement: Impact on Trade, Development and Innovation.

[8] Not all of these factors are necessarily designed purposefully to reduce or complicate trade, but may serve the purpose of protecting consumers or the environment.

The Role of Trade Facilitation and Regulatory Frameworks when Production is Organized in Supply Chains

From a more traditional cross-border perspective, incorporating a GPN perspective in trade facilitation considerably enhances the benefits expected from regulatory reforms and the upgrade of trade related infrastructure. It is especially true for those developing economies that started their export-led industrialization through EPZs and wish to increase the domestic content of their exports. Incorporating more domestic suppliers in the export-processing activities often calls for reducing transportation costs within the economy and facilitating the process of administrative procedures required to clear customs and get refunds for the taxes paid on the exported part of their output.

Reassessing Effective Protection Rates from a 'Trade in Tasks' Perspective

Effective protection rates (EPRs) are the ratio of sectoral value-added at domestic prices (i.e. buying and selling on the domestic market) relative to what the industry would have gained at world prices (i.e., if there were no tariff duty to be paid on final and intermediate goods). Duty rates are either nil or positive. EPRs can be negative when the nominal protection received on the final output of the firm (the tariff paid by competing imports) is smaller than the duties paid on the intermediate inputs required for its production. A high EPR acts as an anti-export bias, since the value-added obtained is much higher on the domestic market than on the international one. Measuring this bias is therefore particularly relevant in global value chain trade, where what is traded is 'value-added' and not final goods. A report by WTO and IDE-JETRO (2011) shows that EPRs declined in East Asia, as part of their export-led strategy.

EPR computation requires two sets of data: tariff schedules (nominal protection per traded goods) and sectoral input-output coefficients. While tariff schedules are easily obtainable, the availability of updated and harmonized input-output tables has been a limit to the computation of this trade-policy indicator. The availability of MRIO approaches removes the principal bottleneck, by providing harmonized input-output tables that facilitate the calculation of EPRs and their inter-country comparison. By linking national tables into a multi-regional framework, MRIO allow also to assess the impact of a change in tariff schedules on the trade partners.

MRIOs and Measuring the International Economic Impact of Trade Shocks (Macroeconomic Shocks' Simulation)

As mentioned, MRIOs enable us to estimate the supply-side transmission channels of trade policy or macro-economic shocks. For example, Escaith et al. (2011) carried out a simulation of the disruptive effect of the March 2011 earthquake and tsunami in Japan. This simulation measures the impact on the production costs of trade partners of a shortage in parts and components exported by Japan. The intensity of the supply shock varies from industry to industry, and for each trade partner is proportional to the role of Japanese suppliers for this particular sector. More generally, intermediate trade flows within GPNs are the main channel of transmission of disruptive supply shocks, like natural disasters or abrupt changes in trade policy. MRIOs make it possible to complement the traditional demand-led analysis of shock transmission with its micro-economic supply-side counterpart.

Trade Barriers and Global Competitiveness

The organization of global supply chains relies on each country or company's comparative advantage. Accounting for the foreign value added content in a nation's exports therefore leads us to review the notion of 'us versus them' and to create a greater awareness of mutual dependency among economies. The competitiveness of countries and their export performance is increasingly dependent on imported inputs. Additionally, domestic value added is not only present in exports but can also be included in imported goods previously exported and further processed (circular trade) in a third country. If 'foreign goods' may actually include a significant proportion of domestic value-added, similarly, so-called 'national products' may be predominantly produced outside the national territory. In addition, products with foreign trademarks may be manufactured in the domestic market. Consequently, protectionist measures like tariff increases, anti-dumping measures or 'buying national' may have counter-productive effects on enterprises or jobs they intend to protect. For example an increase in the cost of imports affects domestic companies involved in global production as well as the functioning and competitiveness of the chain itself. In addition, arbitrary changes in trade policy will increase uncertainty and limit the willingness of investors, both domestic and foreign, to set up long-term agreements.

Conclusion

In today's world, countries are no more trading wine for clothes, as in the famous 19th century model of David Ricardo which specifies comparative (or relative) advantage as source of trade between countries. Nowadays, business models are of a more interdependent character whereby the production is sliced into several stages, often geographically differentiated. One of the most visible outcomes of this new type of trade is the rise of key emerging economies and the shift in economic realities. The internationalization of production processes leads also to an increased inter-dependency, expanded trade ties and a more deeply shared interest in a well-functioning trading system. In such a situation, in which governance becomes a global dimension, national policy makers must take into consideration the complex international implications of their domestic economic and social strategies.

However, the lack of appropriate data has so far limited the possibility of basing this global approach to trade policy on factual statistics. The MRIO data, by providing the means to measure the strength and complexity of the internationalization of supply chains, also allows a better understanding of the inter-industry complementarities and how global production networks create value added across countries.

Further reading

Escaith, H., Keck, A., Nee, C., & Teh, R. (2011, April). Japan's earthquake and tsunami: International trade and global supply chain impacts. *VoxEU* .

Grossman, G., & Rossi-Hansberg, E. (2006). The Rise of Offshoring: It's Not Wine for Cloth Anymore. *Proceedings*, (pp. 59-102).

Porter, M. (1985). Competitive advantage: creating and sustaining superior performance. *New York Free Press* .

Timmer, M. P., Erumban, A. A., Los, B., Stehrer, R., & De Vries, G. (2012). New measures of European competitiveness: a global value chain perspective.

WTO & IDE-JETRO. (2011). *Trade patterns and global supply chains in East Asia: from trade in goods to trade in tasks.* WTO.

WTO. (2011). *World Trade Report 2011: The WTO and preferential trade agreements: From co-existence to coherence.* WTO.

Chapter 23: The Importance of Input-Output Data for the Regional Integration and Sustainable Development of the European Union

Joerg Beutel, Isabelle Rémond-Tiedrez,
and José M. Rueda-Cantuche

Introduction

In recent years the interest in supply, use and input-output tables has seen an impressive resurgence. With the globalization of economic activities, many analysts have rediscovered the great utility of these specific statistics for several purposes of policy advice.

Supply and use tables are an integral part of the European System of Accounts (ESA1995) and play an important role as an integration framework for the national accounts. They constitute the centre piece of the internationally compatible accounting framework for a systematic and detailed description of the economy, its various components on the supply and demand side and its relations to other economies.

A new version of the European System of Accounts was approved in 2011 that will be implemented by the Statistical Offices in 2014. It is fully consistent with the revised world-wide System of National Accounts (SNA 2008). Since 2002 all 27 member countries of the European Union have been requested to submit every year supply and use tables and every five years input-output tables.

Isabelle Rémond-Tiedrez heads a team of experts at Eurostat that is collecting and harmonizing the input-output data for the European Union. The team is supported by Jörg Beutel and José M. Rueda-Cantuche in its effort to compile aggregate supply, use and input-output tables for the European Union (EU27) and the Euro area (EA17).

Besides the official supply, use and input-output tables for member countries the European Commission has recently made a substantial investment in specific input-output data for the analysis of productivity (EUKLEMS), environmental issues (EXIOPOL) and international trade (WIOD).

In 2008 Eurostat published the Eurostat Manual of Supply, Use and Input-Output Tables[1]. The key issue of this manual is to describe how supply and use tables can be compiled from the main statistical sources.

The objective of this chapter is to promote the new database of supply, use and input-output tables for multiregional analysis of the European Union and the evaluation of the European Regional and Cohesion policy. Both policies aim to reduce regional disparities in the European Union.

The main benefits of the new database are the deep disaggregation of industries and the direct reflection of statistical sources and survey results. Supply and use tables provide an ideal framework for checking the consistency of statistical data obtained from different sources. The input-output framework also allows basic economic data to be entered into the system in exactly the same structure in which the basic data can be surveyed and observed.

In the first section we explain the concept of national supply and use and input-output tables. Then we discuss the process of developing aggregate supply and use and input-output tables and environmentally extended input-output tables for the European Union. We present extended supply, use and input-output tables for the European Union, which were aggregated from 27 national supply and use tables of the Member countries. These tables include valuable information on intra and extra European trade, which can be used to assess the European regional policy. Supplementary information on air emissions of industries and private households will open new doors for European environmental policies.

This leads us to the importance of input-output data for sustainable development. In the next section a test for sustainable development is made for the Member countries of the European Union and the fifteenth largest economies of the world, which reflects the change of produced capital, human capital and natural capital. We show how input-output data and a measure of sustainable development can be used for the evaluation of the European regional policy.

In the final section we explain how the new input-output database can be used for a more profound assessment of Structural Fund interventions and regional policies than in the past. Structural Funds and the Cohesion Fund are financial tools set up to implement the Regional and Cohesion policy of the European Union. They aim to reduce regional disparities in terms of income, wealth and opportunities. Europe's poorer regions receive most of the support but all European regions are eligible for funding under the policy's various funds and programs. The annual submission of supply and use tables will enlarge the potential for analysis and follow up of actual interventions. The aid package

---

[1] Eurostat (2008): Eurostat Manual of Supply, Use and Input-Output Tables, Luxembourg. The 2008 release of the Manual was written by Jörg Beutel and edited by Peter Ritzmann (Eurostat).

favouring the least developed regions in the European Union has sometimes been compared to the European Recovery Programme (Marshall Plan), when in the period from 1948 to 1952 Western Europe received 12 billion dollars of aid, a sum that was equivalent to 2.1 percent of the average of the receiver nations' GDP.

Indeed grants of the European Union made available for major convergence areas during the seven-year period from 2007 to 2013 represent a similar magnitude in terms of GDP. In view of the development and structural adjustment needs of the regions whose development is lagging behind, the expenditure volume of convergence interventions is substantial in relation to expected gross domestic product.

Input-Output Framework

The input-output framework of the European System of Accounts (ESA) consists of two types of tables: the supply and use tables and symmetric input-output tables

The input-output tables and in particular the supply and use tables serve statistical and analytical purposes. They provide a framework for checking the consistency of statistics on flows of goods and services obtained from quite different kinds of statistical sources: industrial surveys; household expenditure inquiries; investment surveys; foreign trade statistics; and other statistics.

As an analytical tool input-output data are conveniently integrated into macroeconomic models in order to analyze the links between demand and supply, in particular between final demand components and industrial output levels. Input-output data also serve a number of other analytical purposes by linking other major statistics (employment, capital, energy, and environment) to the system of national accounts.

Supply and use tables and input-output tables provide the basis for many analyses within the European Commission. Detailed and reliable data on the interdependent production process of all Member States will serve the integration process.

National Supply, Use and Input-Output Tables of the European Union

The national supply, use and input-output tables of the European Union comprise specific information on domestic production but also on foreign trade, separating exports and imports to and from members of the European Union, the Euro area and third countries.

Table 1: Supply table at basic prices, including a transformation into purchasers' prices

Germany 2007

No	PRODUCTS	(1) Agriculture, forestry and fishing	(2) Industry incl. energy	(3) Construction	(4) Trade, transport and communication services	(5) Financial services and business activities	(6) Other services	(7) Domestic output at basic prices	(8) Intra EU imports	(9) Extra EU imports	(10) Total imports	(11) Total supply at basic prices	(12) Trade and transport margins	(13) Taxes less subsidies on products	(14) Total supply at purchasers prices
		OUTPUT OF INDUSTRIES							IMPORTS				VALUATION		
1	Products of agriculture, forestry and fishing	50	0				0	50	14	9	23	73	15	4	91
2	Industrial proucts incl. energy	0	1,708	5	9		1	1,723	457	321	777	2,501	342	165	3,008
3	Construction work	0	9	200	2	0	1	213	2	1	3	216		24	240
4	Trade, transport and communication services	0	52	1	717	0	1	772	28	20	48	820	-357	14	477
5	Financial services and business services	0	38	1	8	989	1	1,037	31	24	55	1,092		37	1,129
6	Other services		3	1	0		696	699	3	3	6	705		8	713
7	Total	51	1,810	207	736	991	700	4,495	534	378	912	5,407	252	252	5,659
8	Cif/ fob adjustments on imports								-1	-2	-2	-2			-2
9	Direct purchases abroad by residents								38	20	58	58			58
10	Total	51	1,810	207	736	991	700	4,495	572	396	968	5,463		252	5,714
	TOTAL OF WHICH														
1	Market output	49	1,801	186	727	852	341	3,956							
2	Output for own final use	2	8	22	6	128	12	178							
3	Other non-market output				3	11	347	361							

Table 2: Use table at purchasers' prices

Germany 2007		INPUT OF INDUSTRIES						CONSUMPTION			CAPITAL FORMATION		EXPORTS			
		Agriculture, forestry and fishing	Industry incl. energy	Construction	Trade, transport and communication services	Financial services and business activities	Other services	Total	Final consumption expenditure by households	Final consumption expenditure by non-profit organisatio	Final consumption expenditure by government	Gross fixed capital formation	Changes in inventories	Exports intra EU	Exports extra EU	Final uses at purchasers prices	Total use at purchasers basic prices
No	PRODUCTS	1	2	3	4	5	6	7	8	9	10	11	12	13	14	15	16
1	Products of agriculture, forestry and fishing	3	42	0	2	1	3	50	26			4	3	6	2	42	91
2	Industrial proucts incl. energy	17	898	74	85	22	70	1,166	624		33	216	-12	632	347	1,842	3,008
3	Construction work	0	7	9	4	21	9	51	4			185		0	0	189	240
4	Trade, transport and communication services	1	63	2	146	10	15	237	173		3		-2	35	31	240	477
5	Financial services and business services	9	198	31	108	275	66	687	325	4	7	46		28	31	442	1,129
6	Other services	1	30	2	13	25	57	128	153	33	392	4	0	1	1	586	713
7	Total at basic prices	31	1,238	119	357	355	219	2,318	1,306	37	436	456	-10	704	412	3,340	5,659
8	Ciff/fob adjustments on exports								58					-1	-2	-2	-2
9	Direct purchases abroad by residents								-26					17	9	58	58
10	Purchases on domestic territory by non-resid																
11	Total at purchasers' prices	31	1,238	119	357	355	219	2,318	1,339	37	436	456	-10	720	420	3,396	5,714
12	Compensation of employees	8	350	52	233	201	335	1,180									
13	Other net taxes on production	-6	10	1	12	18	-6	28									
14	Consumption of fixed capital	8	79	5	47	152	70	360									
15	Operating surplus, net	11	133	30	88	265	81	608									
16	Value added at basic prices	21	572	88	380	636	480	2,177									
17	Output at basic prices	51	1,810	207	736	991	700	4,495									
SUPPLEMENTARY DATA																	
1	Fixed capital formation (bn Euro)	8	83	4	64	228	82	470									
2	Fixed capital stock (bn Euro)	267	1,399	77	992	6,792	2,562	12,090									
3	Labour inputs (Mio. persons)	1	8	2	10	7	12	40									

Table 3: Symmetric input-output table at basic prices

No	Germany 2007 / PRODUCTS	INPUT OF HOMOGENEOUS INDUSTRIES							CONSUMPTION			CAPITAL FORMATION		EXPORTS		Final uses at purch asres' prices	Total use at purch asers' basic prices
		Agric ulture, forest ry and fishin g	Indust ry incl. energ y	Const ructio n	Trade , transp ort and comm unicat ion servic es	Finan cial servic es and busin ess activiti es	Other servic es	Total	Final consu mptio n expen diture by house holds	Final consu mptio n expen diture by non-profit organ isatio	Final consu mptio n expen diture by gover nment	Gross fixed capita l forma tion	Chan ges in invent ories	Expor ts intra EU	Expor ts extra EU		
		1	2	3	4	5	6	7	8	9	10	11	12	13	14	15	16
1	Products of agriculture, forestry and fishing	2	26		0	1	1	30	9			3	2	4	1	20	50
2	Industrial proucts incl. energy	9	474	49	46	14	32	624	228		8	117	-15	476	285	1,099	1,723
3	Construction work	0	6	7	4	21	8	45	4			164		0	0	167	213
4	Trade, transport and communication services	3	106	13	137	10	22	292	334		16	22	-2	65	46	481	772
5	Financial services and business services	8	170	32	109	241	55	615	315	4	7	36		28	31	422	1,037
6	Other services	1	26	2	13	27	51	121	147	33	392	4	0	1	1	579	699
7	Total at basic prices	24	808	104	309	313	170	1,727	1,037	37	423	346	-15	575	365	2,768	4,495
8	Use of imported products, cif	5	386	17	53	39	26	526	125		7	74	5	129	47	386	912
9	Taxes less subsidies on products	1	13	2	12	15	23	65	145		6	36		0	0	186	252
10	Total at purchasers' prices	30	1,206	122	373	368	219	2,318	1,306	37	436	456	-10	704	412	3,340	5,659
11	Compensation of employees	8	326	53	249	210	334	1,180									
12	Other net taxes on production	-6	10	1	12	18	-6	28									
13	Consumption of fixed capital	7	71	5	46	160	70	360									
14	Operating surplus, net	11	111	32	92	281	82	608									
15	Value added at basic prices	20	517	90	399	669	481	2,177									
16	Output at basic prices	50	1,723	213	772	1,037	699	4,495									

A *supply table* shows the supply of goods and services by product and by type of supplier, distinguishing output by domestic industries and imports. In the official transmission programme for the European system of national and regional accounts the supply table is given at basic prices (Table 1) including a transformation into purchasers' prices. The primary activities of industries are shown on the diagonal of the supply table and the secondary activities off the diagonal.

A *use table* shows the use of goods and services by product and by type of use (Table 2). Furthermore the table shows the components of value added by industry. A special feature of the use table is supplementary data for each industry on fixed capital formation, fixed capital stock and labour input. Consequently the columns of the use table reflect for each sector all inputs that are required for production. Listed are intermediate inputs and primary inputs (capital, labour, land, natural resources).

A symmetric input-output table (Table 3) is a matrix describing the domestic production processes and the transactions in products of the national economy in great detail. A symmetric input-output table rearranges both supply and use in a single table.

There is one major conceptual difference between a symmetric input-output table and supply and use tables. Supply and use tables relate products to industries, while a symmetric input-output table relates products to products or industries to industries. So, in a symmetric input-output table either a product or an industry classification is employed for both rows and columns.

Supply, Use and Input-Output Tables for the European Union and the Euro Area

The project of creating consolidated and aggregated supply, use and input-output tables for the European Union (EU27) has been conducted by Eurostat with support of the Joint Research Centre's Institute for Prospective Technological Studies (JRC-IPTS) and the Konstanz University of Applied Sciences in Germany. That work has focussed on creating consolidated Supply and Use Tables (SUT) for the aggregated European Union (EU27) and the euro area (EA17) for 2000 to 2007.

The compilation of aggregate supply, use and input-output tables for the European Union requires the following annual information for each member country:

1) Supply table at basic prices, including a transformation to purchasers' prices
2) Use table at purchasers' prices
3) Valuation matrix for trade and transport margins
4) Valuation matrix for taxes less subsidies on products
5) Supply table at basic prices
6) Use table at basic prices
7) Use table for domestic output at basic prices
8) Use table of imports at basic prices

For the Statistical Offices it is only mandatory to submit every year the first two sets of tables to Eurostat. Therefore, Eurostat approached the Statistical Offices to provide the valuation matrices and supply and use tables at basic prices on a voluntary basis. In Table 4 the individual supply tables at basic prices of the 27 individual countries have been aggregated for the European Union (EU27). The aggregated supply table includes separate vectors for intra EU imports and extra EU imports.

In Table 5 the individual use tables of domestic output have been aggregated with separate matrices for intra EU imports (rows 8-13) and extra EU imports (rows 15-20). In addition the aggregate table includes separate columns for intra EU exports and extra EU exports.

The methodology for the compilation of consolidated and aggregate supply, use and input-output tables is explained in a technical documentation of Eurostat.[2] The consolidated tables are based on balanced intra-European trade for the European Union and the Euro area. This information offers various options for regional analysis.

Table 4: Simple aggregation of supply tables at basic prices for the European Union

European Union 2006	OUTPUT OF INDUSTRIES (NACE)						Domestic output at basic prices	IMPORTS			Total supply at basic prices
	Agriculture, forestry and fishing	Industry incl. energy	Construction	Trade, transport and communication services	Financial services and business activities	Other services		Intra EU imports	Extra EU imports	Total	
No PRODUCTS (CPA)	1	2	3	4	5	6	7	8	9	10	11
1 Products of agriculture, forestry and fishing	377	2	0	2	0	1	382	57	40	97	479
2 Industrial proucts incl. energy	15	6,727	11	78	12	9	6,852	2,198	1,406	3,604	10,456
3 Construction work	2	20	1,663	15	15	5	1,720	5	3	8	1,728
4 Trade, transport and communication services	4	174	8	4,370	24	16	4,597	154	96	250	4,847
5 Financial services and business services	2	147	23	97	4,774	70	5,112	239	145	384	5,496
6 Other services	2	20	1	17	7	3,529	3,578	25	21	46	3,623
7 Total	403	7,089	1,707	4,579	4,832	3,631	22,241	2,678	1,711	4,390	26,630
TOTAL OF WHICH											
1 Market output	299	5,552	1,322	3,531	3,130	690	14,524				
2 Output for own final use	20	37	60	16	727	67	927				
3 Other non-market output	83	1,500	325	1,032	976	2,874	6,790				

---

[2] Eurostat (2011): Creating consolidated and aggregated EU27 Supply, Use and Input-Output Tables, adding environmental extensions (air emissions) and conducting Leontief-type modelling to approximate carbon and other footprints' of EU27 consumption for 2000 to 2006, Luxembourg.

Table 5: Simple aggregation of use tables at basic prices for the European Union

No	PRODUCTS (CPA)	Agriculture, forestry and fishing [1]	Industry incl. energy [2]	Construction [3]	Trade, transport and communication services [4]	Financial and business services [5]	Other services [6]	Total [7]	Final consumption expenditure by households [8]	Final consumption expenditure by non-profit organisatio [9]	Final consumption expenditure by government [10]	Gross fixed capital formation [11]	Changes in inventories [12]	Exports intra EU [13]	Exports extra EU [14]	Final uses at purchasers' prices [15]	Total use at purchasers' basic prices [16]
1	Products of agriculture, forestry and fishing	46	160	.	18	1	5	233	88	0	2	6	5	37	12	149	382
2	Industrial proucts incl. energy	69	1,864	305	408	132	207	2,986	1,115	0	37	382	8	1,515	809	3,866	6,852
3	Construction work	3	41	344	44	109	46	586	45	0	3	1,075	4	6	2	1,134	1,720
4	Trade, transport and communication services	32	604	104	749	194	167	1,850	1,932	0	66	141	0	383	225	2,747	4,597
5	Financial services and business services	25	609	167	705	1,137	345	2,987	1,388	6	53	263	4	233	179	2,126	5,112
6	Other services	5	78	9	66	93	261	512	639	154	2,213	21	1	21	16	3,066	3,578
7	Total domestic products at basic prices	180	3,356	931	1,989	1,666	1,031	9,153	5,208	161	2,373	1,888	21	2,195	1,243	13,088	22,241
8	Products of agriculture, forestry and fishing	4	24	2	2	0	1	31	17	0	0	1	1	6	1	26	57
9	Industrial proucts incl. energy	16	809	62	95	28	59	1,067	369	0	19	204	9	371	158	1,131	2,198
10	Construction work	0	0	2	0	0	0	3	0	0	0	2	0	0	0	2	5
11	Trade, transport and communication services	1	26	2	58	10	6	104	28	0	0	1	0	13	8	50	154
12	Financial services and business services	1	49	8	32	85	15	190	10	0	2	11	0	17	9	49	239
13	Other services	0	3	0	3	1	6	13	5	0	2	1	0	2	1	12	25
14	Total imports from EU countries	22	912	75	189	125	87	1,409	429	1	23	221	10	409	177	1,289	2,678
15	Products of agriculture, forestry and fishing	3	17	0	2	0	1	22	12	0	0	1	0	4	1	18	40
16	Industrial proucts incl. energy	7	575	29	49	16	37	713	216	0	9	120	12	232	104	693	1,406
17	Construction work	0	0	1	0	0	0	2	0	0	0	1	0	0	0	1	3
18	Trade, transport and communication services	0	14	1	42	7	4	67	16	0	0	0	0	7	5	29	96
19	Financial services and business services	1	32	5	20	48	11	116	7	0	1	0	0	9	6	29	145
20	Other services	0	3	0	2	1	6	11	4	0	2	1	0	2	0	10	21
21	Total imports from third countries	11	640	37	115	72	58	932	255	1	12	130	12	253	117	779	1,711
22	Total products at basic prices	212	4,908	1,043	2,293	1,862	1,177	11,495	5,891	162	2,409	2,238	43	2,856	1,537	15,136	26,630
23	Taxes less subsidies on products	5	68	17	78	77	96	340	737	0	6	178	2	8	6	937	1,277
24	Total products at purchasers' prices	217	4,976	1,060	2,371	1,939	1,272	11,835	6,628	162	2,415	2,417	44	2,864	1,543	16,073	27,908
25	Compensation of employees	65	1,210	385	1,292	1,158	1,780	5,890									
26	Other net taxes on production	-22	126	30	135	118	144	532									
27	Consumption of fixed capital	21	205	22	163	428	150	989									
28	Operating surplus, net	122	572	210	619	1,189	283	2,995									
29	Value added at basic prices	186	2,113	648	2,208	2,893	2,358	10,406									
30	Output at basic prices	403	7,089	1,707	4,579	4,832	3,631	22,241									

Extended Input-Output Tables for the European Union

The objective of the first project was to compile consolidated input-output tables for the European Union. The second project focuses on creating and analysing European environmentally extended input-output tables. This project was commissioned by Eurostat and undertaken by a consortium consisting of the Netherlands Organisation for Applied Scientific Research (TNO), the Centre of Environmental Sciences of Leiden University (CML), the Norwegian University of Science and Technology (NTNU), and University of Groningen (RUG). The extended input-output table for the European Union includes satellite information on air emissions by industries and private households in 1.000 tons (Table 6).

Table 6: Extended input-output table at basic prices for the European Union

European Union 2006 (values in 1.000 tons for EXTENSIONS)

No	PRODUCTS (CPA)	1 Agriculture, forestry and fishing	2 Industry incl. energy	3 Construction work	4 Trade, transport and communication services	5 Financial service and business activities	6 Other services	7 Total	8 Final consumption expenditure by households	9 Final consumption expenditure by non-profit organisations servi	10 Final consumption expenditure by government	11 Gross fixed capital formation	12 Changes in inventories and valuables	13 Exports intra EU	14 Exports extra EU	15 Final uses at purchasers' prices	16 Total use at basic purchasers' prices
1	Products of agriculture, forestry and fishing	46	173	3	24	4	6	255	100	0	2	7	5		13	127	382
2	Industrial proucts incl. energy	77	2,421	359	520	219	257	3,852	1,418	0	51	546	16		968	2,999	6,852
3	Construction work	3	42	339	45	115	46	590	45	0	3	1,077	4		2	1,131	1,720
4	Trade, transport and communication	35	722	117	828	251	179	2,132	2,014	0	68	150	0		233	2,465	4,597
5	Financial services and business ervices	25	650	178	722	1,248	357	3,180	1,402	6	54	279	4		188	1,933	5,112
6	Other services	5	78	10	69	103	258	523	643	155	2,216	22	1		17	3,054	3,578
7	Total domestic products at basic prices	191	4,087	1,005	2,207	1,941	1,102	10,532	5,621	161	2,396	2,081	30		1,420	11,709	22,241
8	Imported products	11	635	38	129	86	63	963	270	1	13	157	13		117	571	1,534
9	Total at basic prices	201	4,722	1,043	2,336	2,027	1,165	11,495	5,891	162	2,409	2,238	43		1,537	12,280	23,774
10	Taxes less subsidies on products	5	66	17	77	82	93	340	737	0	6	178	2		6	929	1,270
11	Total at purchasers' prices	206	4,788	1,061	2,413	2,109	1,259	11,835	6,628	162	2,415	2,417	44		1,543	13,209	25,044
12	Compensation of employees	63	1,183	387	1,277	1,232	1,748	5,890									
13	Other net taxes on production	-21	122	31	133	124	142	532									
14	Consumption of fixed capital	20	198	26	163	436	147	989									
15	Operating surplus, net	114	561	216	610	1,212	281	2,995									
16	Value added at basic prices	175	2,064	660	2,184	3,003	2,319	10,406									
17	Output at basic prices	382	6,852	1,720	4,597	5,112	3,578	22,241									
	EXTENSIONS																
1	Sulphur oxides (SO2)	124	6,687	35	1,326	18	100	8,289	487							487	8,776
2	Nitrogen oxides(NO2)	1,134	4,396	441	5,201	248	365	11,774	2,156							2,156	13,930
3	Ammonia(NH3)	3,612	71	1	8	2	73	3,768	107							107	3,874
4	Carbon monoxide (CO)	2,259	7,905	378	2,273	335	645	13,796	16,161							16,161	29,957
5	Non-methane volatile organic compounds (VOC)	1,943	3,400	583	785	76	278	7,065	3,213							3,213	10,278
6	Methane (CH4)	9,741	3,616	7	196	9	5,611	19,178	638							638	19,816
7	Nitrous oxide (N2O)	864	205	3	21	1	55	1,149	34							34	1,184
8	Carbon dioxide (CO2)	91,589	2,559,801	63,896	607,430	65,693	137,155	3,525,564	923,670							923,670	4,449,234

The matrix below the input-output table is a satellite system on air emissions which has been integrated into the input-output framework. This system provides useful information on the emission of industries and private households in form of sulphur dioxide (SO_2), Nitrogen oxides (NOx), Ammonia (NH_3), Carbon monoxide (CO), non-methane volatile organic compounds (NMVOC), methane (CH_4), nitrous oxide (N_2O), and carbon dioxide (CO_2).

This information allows us to assess the emission of greenhouse gases and to evaluate various environmental policies at the regional level but also at the aggregate level of the European Union.

Sustainable development

In this section we discuss the importance of input-output data to evaluate the sustainable development of nations.

Many economists argue that the maintenance of sustainable economic growth over the long run merely depends on the adequacy of investment expenditure for the various forms of capital. A key assumption of this position is that there will continue to be a high degree of substitution between the different forms of physical, human and natural capital. Under this interpretation of sustainability, it is not necessary to single out natural capital for special treatment. It is simply regarded as another form of capital.

For sustainable development all that is required is that we pass on an aggregate capital stock of produced capital (buildings, machinery, and transport equipment), human capital (education, knowledge, skills) and natural capital (natural resources, ecosystems) that is not smaller than the one that exists now. This view allows passing on to the next generation fewer natural resources as long as we offset the loss by increasing the stock of man-made capital and human capital.

The capital stock of produced capital grows if gross fixed capital formation (new buildings and machinery) is larger than the consumption of produced fixed capital (retirements of old buildings and machinery). The capital stock can only grow if net investment is positive.

Gross national saving is defined as gross fixed capital formation plus current account surplus/deficit (net exports of goods and services + net factor income from abroad). Net national saving is the sum of gross national saving less consumption of fixed produced capital.

Adjusted net saving (genuine saving) measures the true rate of savings in an economy after taking into account investment in human capital, depletion of natural resources and damage caused by pollution (Figure 1).

Figure 1: Measuring sustainable development

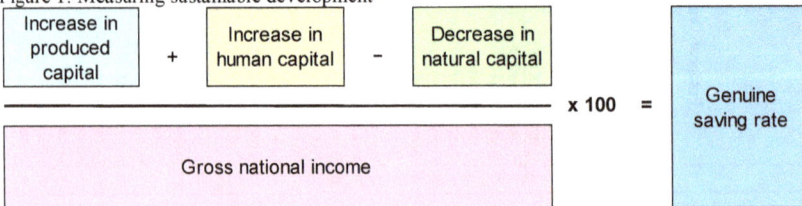

$$\frac{\text{Increase in produced capital} + \text{Increase in human capital} - \text{Decrease in natural capital}}{\text{Gross national income}} \times 100 = \text{Genuine saving rate}$$

Since 1999 a time series of genuine saving or adjusted net saving has been estimated and published by the World Bank in the sequence of World Development Indicators of the World Bank. The definition of the published indicator is as follows:

Gross capital formation
+ Net exports of goods and services
+ Net income from abroad
= Gross national saving
- Consumption of fixed capital
= Net domestic saving
+ Education expenditure
 - Energy depletion
 - Mineral depletion
 - Net forest depletion
 - Damage from carbon dioxide emissions
 - Damage from particulate emissions
= Adjusted net saving (genuine saving)

Adjusted net saving is a national accounting aggregate designed to measure the net change in assets in a national balance sheet that includes natural and human capital. There is an intrinsic linkage between change in wealth of a nation and the sustainability of a development path. If genuine saving is negative at a certain point in time, then future welfare will be less than current welfare. Therefore, adjusted net saving can be regarded as a sustainability indicator. An economy is sustainable if it saves more than the depreciation on its man-made and natural capital.

In Table 7 an assessment has been made which is based on a long time series for adjusted net savings of the World Bank[3] for 210 countries of the world. The sustainability of economic development of a nation is endangered if the sustainability indicator is negative.

---

[3] World Bank (2012): Adjusted net saving time series by country 1970-2008, Washington D.C.

Table 7: Test for sustainable development for the European Union

Country	Gross national saving (1)	Consumption of fixed capital (2)	Net national saving (3)	Education expenditure (4)	Energy depletion (5)	Mineral depletion (6)	Net forest depletion (7)	Carbon dioxide emissions damage (8)	Particulate emissions damage (9)	Adjusted net saving (10)
Austria	27.2	14.3	12.9	5.3	0.2	0.0	0.0	0.1	0.1	17.6
Belgium	25.4	13.9	11.4	5.8	0.0	0.0	0.0	0.2	0.1	17.0
Bulgaria	14.1	11.6	2.5	4.1	1.1	0.8	0.0	0.9	0.9	2.9
Cyprus	5.6	14.4	-8.8	6.5	0.0	0.0	0.0	0.3	0.3	-2.8
Czech Republic	24.2	13.8	10.4	4.4	0.7	0.0	0.0	0.5	0.0	13.4
Germany	25.4	13.8	11.6	4.3	0.3	0.0	0.0	0.2	0.0	15.4
Denmark	23.6	14.2	9.4	7.4	3.0	0.0	0.0	0.1	0.0	13.7
Estonia	20.1	13.5	6.6	4.6	1.5	0.0	0.0	0.7	0.0	9.0
Spain	20.6	14.0	6.6	3.9	0.0	0.0	0.0	0.2	0.2	10.1
Finland	24.8	14.1	10.7	5.6	0.0	0.1	0.0	0.2	0.0	16.0
France	18.7	13.9	4.9	5.1	0.0	0.0	0.0	0.1	0.0	9.8
Greece	7.4	13.9	-6.5	2.8	0.3	0.1	0.0	0.2	0.3	-4.8
Hungary	15.9	15.1	0.8	5.3	0.8	0.0	0.0	0.3	0.0	5.0
Ireland	19.7	17.1	2.5	5.2	0.0	0.0	0.0	0.1	0.0	7.5
Italy	18.5	14.0	4.5	4.5	0.2	0.0	0.0	0.2	0.1	8.5
Lithuania	15.2	12.7	2.5	4.6	0.1	0.0	0.1	0.3	0.1	6.6
Luxembourg	35.8	18.7	17.1	3.7	0.0	0.0	0.0	0.2	0.0	20.6
Latvia	22.3	12.6	9.6	5.6	0.0	0.0	0.2	0.2	0.0	14.8
Malta	16.0	13.8	2.2	4.6	0.0	0.0	0.0	0.3	0.0	6.6
Netherlands	10.3	13.9	-3.6	4.8	2.0	0.0	0.0	0.2	0.2	-1.2
Poland	19.1	12.7	6.4	5.4	1.5	0.3	0.1	0.5	0.2	9.2
Portugal	12.6	13.6	-1.0	5.3	0.0	0.1	0.0	0.2	0.0	4.1
Romania	25.0	11.7	13.3	3.4	2.4	0.1	0.0	0.4	0.0	13.7
Sweden	27.1	12.5	14.6	6.4	0.0	0.4	0.0	0.1	0.0	20.5
Slovenia	27.0	13.6	13.4	5.3	0.1	0.0	0.2	0.2	0.1	18.1
Slovakia	21.9	13.1	8.8	3.7	0.1	0.0	0.4	0.4	0.0	11.7
United Kingdom	14.8	13.7	1.2	5.1	2.1	0.0	0.0	0.2	0.0	3.9

% of Gross national income 2008 (GNI)

In 2008 low sustainability indicators are reported for several member countries of the European Union: Greece (-4.8%), Cyprus (-2.8%), Netherlands (-1.2 %), Bulgaria +2.9%) and United Kingdom (+3.9%). High ratios are listed for Luxembourg (+20.6%), Sweden (+20.5%), Slovenia (+18.1%), Austria (+17.6%) and Belgium (+17.0%).

The difference between the United States (+0.9%) and the European Union (+ 12.8%) or Japan (+15.3 %) in 2008 is substantial for such large highly developed economies (Table 8).

Table 8: Test for sustainable development for large economies

Country	Gross national saving (1)	Consumption of fixed capital (2)	Net national saving (3)	Education expenditure (4)	Energy depletion (5)	Mineral depletion (6)	Net forest depletion (7)	Carbon dioxide emissions damage (8)	Particulate emissions damage (9)	Adjusted net saving (10)
	% of Gross national income 2008 (GNI)									
Euro area	22.6	14.0	8.6	4.6	0.3	0.0	0.0	0.2	0.0	12.8
United States	12.6	14.0	-1.4	4.8	1.9	0.1	0.0	0.3	0.1	0.9
China	53.9	10.1	43.8	1.8	6.7	1.7	0.0	1.3	0.8	35.1
Japan	25.9	13.3	12.6	3.2	0.0	0.0	0.0	0.2	0.3	15.3
Brazil	17.5	11.8	5.8	4.8	2.7	2.3	0.0	0.2	0.1	5.2
India	38.2	8.5	29.7	3.2	4.9	1.4	0.8	1.2	0.5	24.2
Canada	23.4	14.0	9.4	4.8	5.5	0.6	0.0	0.3	0.1	7.6
Russia	32.8	12.4	20.4	3.5	20.5	1.0	0.0	0.9	0.1	1.5
Mexico	25.3	12.0	13.3	4.8	8.2	0.3	0.0	0.3	0.3	9.0
South Korea	30.5	12.6	17.9	3.9	0.0	0.0	0.0	0.4	0.3	21.1
Australia	32.9	14.7	18.1	5.1	4.1	3.8	0.0	0.3	0.0	15.0
Turkey	17.7	11.8	5.9	3.7	0.3	0.1	0.0	0.3	0.6	8.3
Indonesia	22.2	10.7	11.6	1.1	12.6	1.4	0.0	0.6	0.5	-2.4
Switzerland	33.5	13.3	20.2	4.7	0.0	0.0	0.0	0.1	0.1	24.7
Saudi Arabia	48.3	12.5	35.9	7.2	43.5	0.0	0.0	0.6	0.7	-1.8

World Bank: Adjusted net saving time series by country 1970-2008, Washington D.C.

It remains to be seen if in the foreseeable future we can observe more convergence between the large competitive regions of the world. For the European Union the prime objectives of regional policies are to reduce the development gap and to enhance sustainable development for all member countries.

The input-output framework of the European System of Accounts allows establishing an assessment of sustainable development if satellite systems for produced capital, human capital and natural resources are linked to the system. Some of the required information can be directly derived from the supply and use tables.

In the next section it will be shown how input-output data can be used to evaluate the European regional policy. It will also comprise a policy towards sustainable development of the European Union.

Evaluation of European Regional Policy with Input-Output Data

The key question in this section is how input-output data and information on sustainable development can be used for the evaluation of the European regional policy.

Input-output tables with detailed information for many industries constitute the best database to quantify the economic impacts of regional policies of the European Commission.

Jörg Beutel has evaluated the economic impacts of the Structural Funds for the European Commission several times. The purpose of the 2002 study was to quantify the economic impacts of Objective 1 interventions of the Structural Funds on the basis of the latest input-output tables and macroeconomic data.[4] Objective 1 was defined as promoting the development and structural adjustment of the regions whose development is lagging behind.

The Structural Funds are made up of the European Regional Development Fund (ERDF), the European Social Fund (ESF) and the Cohesion Fund. Together with the European Agricultural Guarantee Fund (EAGF), the European Agricultural Fund for Rural development (EAFRD) and the European Fisheries Fund (EFF) they have the potential to contribute to regional development and make up the majority of total EU spending.

For 2007-13, the three main objectives of the EU's cohesion policy are: Convergence, Regional Competitiveness and Employment, and European Territorial Cooperation. The aim of the convergence objective is to reduce regional disparities in Europe by helping regions to catch up with the ones better off. For the Structural Funds (ERDF and ESF) regions are eligible where per capita GDP is below 75% of the European average. Member states are eligible for the Cohesion Fund whose per capita gross national income (GNI) is below 90% of the European average.

In order to evaluate the economic impacts of Structural Fund interventions an analysis system was developed for the Directorate-General for Regional Policies based on harmonized input-output tables for the European Union and a common methodology for impact analysis for all member countries. With a new set of annual supply and use tables comprising labour and capital stock data, Eurostat

---

[4] Beutel, Jörg (2002): The economic impact of objective 1 interventions for the period 2000-2006, Report to the Directorate-General for Regional Policies, Konstanz.

provides the appropriate database for the assessment of Structural Fund interventions.

In the past the main task of the impact analysis system was to quantify how much of expected development could be attributed to Objective 1 expenditures for regions whose development was lagging behind. Separate analysis was conducted for three types of expenditures:

- Interventions of the European Union (Structural Funds)
- Public interventions (Structural Funds, national public interventions)
- Total interventions (Structural Funds, national public interventions, national private participation).

Figure 2: Eligible areas in the EU under the Convergence Objective

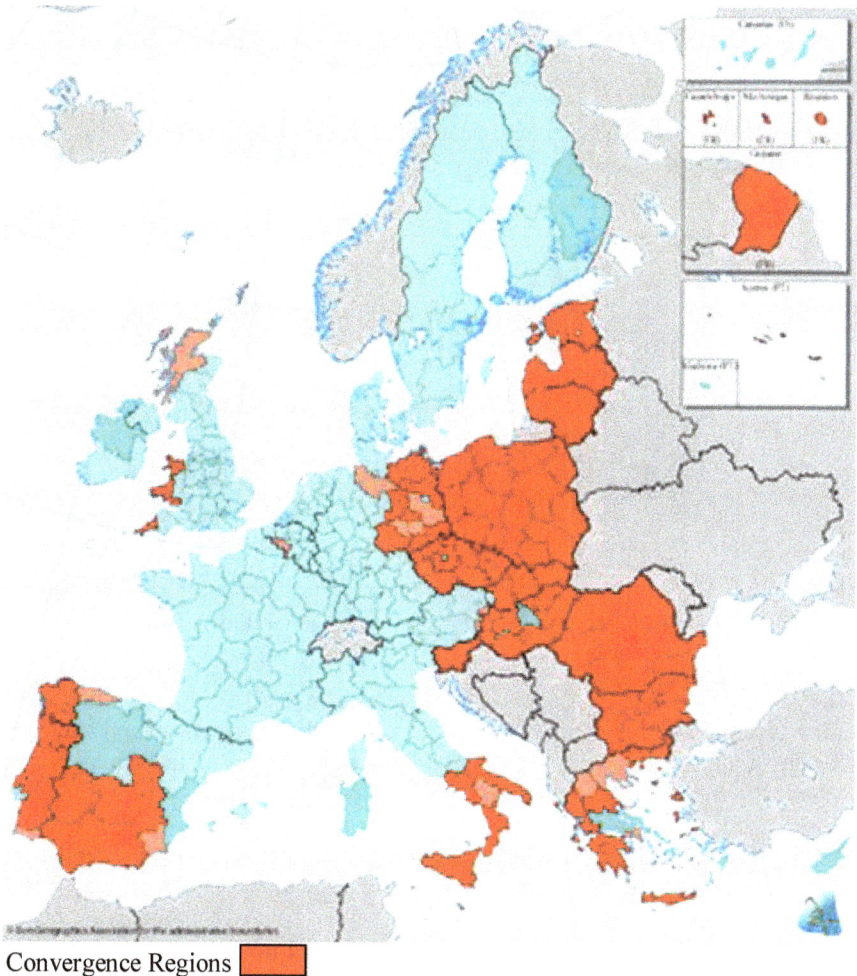

Convergence Regions

Presently, areas of the European Union qualify as eligible under the Convergence Objective (formerly Objective 1) whose per capita gross domestic product measured in purchasing power parities (PPS) are less than 75 % of the European Union's average (Figure 2).

The development gap of convergence regions in the European Union is significant in several member countries (Figure 3). In 2010 the following member countries were eligible under the Convergence Objective: Slovakia (73%), Hungary (65%), Estonia 64%), Poland (63%), Lithuania (57%), Latvia (51%), Romania 47%) and Bulgaria (44%).

Figure 3: GDP per inhabitant in member countries 2010

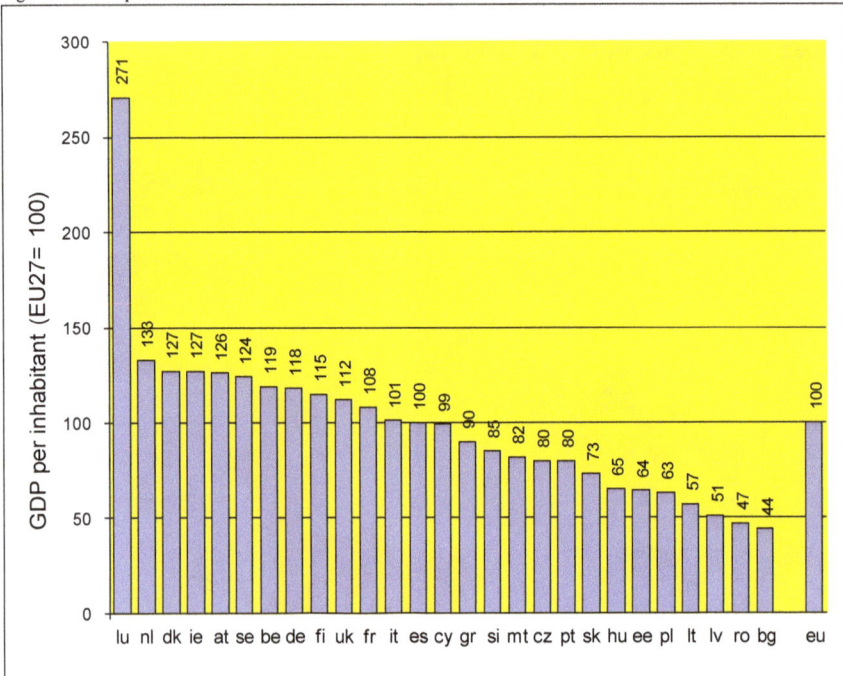

Note: GDP based on purchasing power parities per inhabitant

The main task of the impact analysis system is to analyse how far effects and impacts of the Structural Fund interventions affect the development and structural change of the target regions. The objective is to find comparable answers for the beneficiary Member States on the following main questions:

- How much of the expected economic growth can be attributed to the convergence interventions in general and to interventions of the European Union in particular?
- How will the convergence interventions and the grants of the European Union influence the economic aggregates and the structure of the beneficiary economies? In particular, what part of the grants of the

European Union will be transformed into demand and production in the target region?

- How big a share of the interventions will leak to more prosperous regions via increased demand for imports?
- How can we assess the employment effects of the implementation of the priorities agreed for convergence interventions, i.e. how many jobs depend upon the achievement of the actions of the Objective 1 interventions, and more particularly upon the envisaged financial transfers from the European Union?
- How is the capital stock affected by the convergence interventions?

In a time of structural change and innovation a macroeconomic analysis without sectoral disaggregation would only allow the study of a few and possibly less important impacts of the structural interventions. In this case, the analysis would remain cursory and potentially misleading. The quantification of various structural effects is the main target of the analysis. Input-output analysis is an ideal tool for such disaggregate analysis.

With this impact analysis system, a valuable instrument has been established for an assessment of the economic effects of Structural Funds intervention. The dynamic input-output model software encompasses impact analysis, follow-up and update of the Communities' structural and regional operations. The analysis focuses on the global economic impacts of Community assisted operations on economic variables such as growth, employment, capital use and leakage effects through trade.

The dynamic input-output model has the capacity to evaluate the long-term supply and demand effects of the Community structural policies. Expenditure of the Structural Funds affects the structure and level of final demand but also induces changes in technology, imports, labour and capital use. In particular the long-term effects on capital and labour, output and productivity are the focus of interest. All projections of supply and use tables and input-output tables are based on the European System of Accounts and the latest economic forecasts of the Directorate General of Economic and Financial Affairs of the European Commission.

For each macroeconomic model the financial programming of Structural Fund interventions has been transformed into macroeconomic variables. This transformation is done at the lowest disaggregate level of expenditure. For each investment project it was identified how much was spent on investment (gross fixed capital formation), materials and supplies for operations and maintenance (intermediates) and primary inputs (value added). Some of the expenditure was investment in man-made capital, mainly for new infrastructure of the productive environment, human infrastructure or basic infrastructure. Other expenditures for productive environment and human resources were more related to costs for operations and maintenance of new initiatives which induce purchases of materials and supplies and primary inputs (value added).

The dynamic input-output model was designed in line with the multiplier-accelerator analysis of macroeconomic theory. According to this theory it is expected that new capacities are required if final demand components are growing. In a first round, expenditures of the Structural funds create higher income and an increase in purchasing power. This may lead to further growth of the economy through the stimulation of consumer incomes and purchases. As soon as the production capacity is fully used more private investment is required in a second round to enlarge the production capacity to meet the growing demand.

In the first part of the model, an estimation is made of how the economy is affected by an increase of gross fixed capital formation which was financed by the Structural Funds to improve the infrastructure of public and private institutions. In the second part, the analysis identifies the mechanism how the contributions of the European Union affect value added in the different industries and generate economic growth. In the third part of the impact analysis system, a dynamic version of the input-output model is used to evaluate the long-term supply effects of the Structural Funds on employment, structural change and economic growth.

However, a dynamic input-output model is not the only choice for impact analysis of the Structural Funds from a macroeconomic perspective. It is far more important that Eurostat's annual supply and use tables provide a new macroeconomic data base which can be used for any type of macroeconomic model for the evaluation of the Structural Funds with a minimum of sectoral disaggregation.

A special feature of the new database is the capability to assess the leakage effects of the structural interventions resulting from intra-European trade. If for example a new power plant is built in southern Italy with European funding it is likely that the region around Milano will supply most of the equipment. In consequence the more prosperous region in northern Italy will benefit more from the project than southern Italy. A set of import matrices covering intra EU imports and extra EU imports as well as the information on intra and extra EU exports enable an assessment of such leakage effects for the European Union.

Conclusion

From a wider perspective of a territory the annual supply and use tables of all Member States of the European Union can also be seen as a set of regional input-output tables for the European Union. This valuable database can be used to compile aggregate input-output tables for the European Union. These data also offer new opportunities for the evaluation of regional policies in various areas of the EU. For the European regional policy it is a prime objective to support regions whose development is lagging behind. It is a clear policy of the European Commission to work towards the regional integration of Europe and its sustainable development. The evaluation of GDP per capita and genuine savings ratios for the Member States clearly indicates that there is still a long way to go.

Further reading

Beutel, J. (2004). Supply and use tables - A new data base for the impact analysis of the Structural Funds, Fifth European Conference on Evaluation of the

Structural Funds, Budapest, 26/27 June 2003. *Challenges for Evaluation in an Enlarged Europe.* Brussels: European Commission.
http://ec.europa.eu/regional_policy/sources/docconf/budapeval/index_en.htm

Beutel, J. (2002). *The economic impact of objective 1 interventions for the period 2000-2006, Report to the Directorate-General for Regional Policies.* Regional Policies, Konstanz.
http://ec.europa.eu/regional_policy/sources/docgener/studies/study_en.htm

Eurostat. (2011). *Creating consolidated and aggregated EU27 Supply, Use and Input-Output Tables, adding environmental extensions (air emissions) and conducting Leontief-type modelling to approximate carbon and other footprints' of EU27 consumption for 2000 to 2006.* Luxembourg: Eurostat.
http://epp.eurostat.ec.europa.eu/portal/pls/portal/!PORTAL.wwpob_page.show?_docname=2530266.PDF

Eurostat. (2008). *Eurostat Manual of Supply, Use and Input-Output Tables.* Luxembourg: Eurostat.
http://epp.eurostat.ec.europa.eu/portal/page/portal/product_details/publication?p_product_code=KS-RA-07-013

Rueda-Cantuche, J. M., Beutel, J., Neuwahl, F., Mongelli, I., & Löschel, A. (2009). A Symmetric Input-Output Table for EU27: Latest Progress. *Economic Systems Research, 21*, 59-79.

World Bank. (2012). *Adjusted net saving time series by country 1970-2008.* Washington, D.C.: World Bank.

Chapter 24: How MRIO Can Help APEC to Address its Environmental Objectives

Kirill Muradov and Akhmad Bayhaqi[1]

Introduction

Asia-Pacific Economic Cooperation (APEC) is an international organization focusing on trade and investment liberalization as well as economic cooperation among its member economies. APEC long-term vision of a free and open trade in the Asia Pacific region is supported by various activities implemented by working groups and committees under the directions and guidance of APEC's senior officials.

In contrast to most multilateral and regional organizations that attempt to bind the members through a negotiation process, APEC members strive to voluntarily adopt best practices, model measures learnt from each other and to raise greater public awareness on trade related issues through wider access to information and research findings.

Having recognised trade and investment liberalization and facilitation as a principal driver for creating growth in the Asia Pacific, APEC member economies share an intention to establish a low carbon society where the pursuit of economic growth is reconciled with protection of the environment. The last 5 years have shaped an APEC Green Growth agenda and members are now exploring clean and sustainable development policies.

APEC Economic Leaders announced a number of actions and initiatives on climate change, energy security and clean development at their 2007 meeting in

---

[1] The views expressed in this chapter are those of the author and do not represent the views of the APEC Secretariat or any APEC member economies

Sydney, Australia. Contribution to the reduction of global greenhouse gas emissions in line with the objectives and principles of the United Nations Framework Convention on Climate Change was central to the Sydney Action Agenda which encouraged co-operative actions on energy efficiency, forest issues, low emission technology and innovation, as well as WTO's work on trade in environmental goods and services (EGS).

At the Yokohama 2010 summit, APEC Leaders endorsed the APEC Growth Strategy to help ensure that regional growth and economic integration are sustainable and widely shared among the peoples of the Asia Pacific. The strategy focused on five desired attributes for economic growth – balanced, inclusive, sustainable, innovative, and secure growth. Under the sustainable (green) growth pillar, APEC members agreed to cooperate to establish low-carbon communities, encourage new green industries and jobs, as well as increase the dissemination and utilization of EGS.

More specifically, the 2010 APEC Energy Ministers' Fukui Declaration on Low Carbon Paths to Energy Security highlighted the importance of energy security, energy efficiency and clean energy supply, including introduction of low-emission power sources, assessment of renewable energy options, smart grid technologies and the Low-Carbon Model Town Project.

The objective of this chapter is to discuss APEC environment related initiatives, especially the Green Growth initiatives, and the possible application of MRIO method to analysing related issues for improved policy making.

APEC Green Growth Initiatives: What Role for MRIO?

Echoing the OECD, WTO and other multilateral organizations, APEC members recognize emerging trends of globalized supply chains and trade in intermediate goods which may have important implications on trade policy making.

Cross-border supply chains quickly emerged as a key business model for an expanding range of product groups thanks to the increased circulation of intermediate goods in international trade. Production itself increasingly represents a network-based process ('made in the world') rather than an activity concentrated in a single location. In a demand-driven economy, consumption of specific products in one location may induce production activities, and therefore emissions and resource use in other locations belonging to the entire supply chain.

As long as an economic production-consumption system is viewed as a combination of supply chains, a sustainability practitioner in an APEC economy or elsewhere may face at least two sets of issues in choosing the correct analytical framework for better policy making:

> (1) **Accounting principle**: For a given geographic unit (such as town, province, island, economy) should we only account for emissions caused by direct energy use in significant sectors, e.g. transportation and manufacturing activities? Or should we account for total emissions for which the populace is responsible because of consumption of final goods and services?

(2) **Supply chain boundary**: For a given reporting unit (company, government entity) or a product should we trace emissions back to direct, tier 1 energy suppliers, e.g. power plants, and tier 2 suppliers, e.g. transportation service providers supplying fuel to power plants? Or should we identify emissions occurring throughout the entire supply chain, with an indefinite number of suppliers, leading to final consumption?

MRIO tables offer a consistent framework that is capable of accounting for all direct and indirect cross-border linkages and spillover (multiplicative) effects at the macroeconomic level. Whereas an Input-Output table represents the entire supply chain network of an economy, MRIO provides understanding of the global supply chain network.

Like any model, MRIO has well known limitations owing to the basic assumptions (linear production function, fixed technology assumption etc.). However, MRIO offers a reasonable trade-off between these limitations and advantages, especially on the simplicity, accessibility and confidence (as the underlying Input-Output data come from official statistical sources). Environmental extensions of MRIO analyses, including consumption based emissions accounting and Life Cycle Assessment (LCA) based on global supply chains, may therefore provide better information to track progress about the implementation of APEC green growth initiatives.

Summary of Selected APEC Green Growth Initiatives

Initiative	Brief description	Key message, relevant to MRIO
Biofuels Initiative (2005)	Studies on biofuel resource potential, cost, infrastructure, and sustainability, to assess the biofuel resources in APEC. Delivered by the APEC Biofuels Task Force under a six-year mandate (2005-2011).	TPTMM-EMM (2011): It is recommended that the EWG and individual APEC economies consider and assess in greater detail the resource and employment potential of second-generation biofuels, while ensuring carbon reduction, stable supply and cost-effectiveness according to life cycle assessments.
Environmental Goods and Services Program Framework (2008)	Intended to support the development of the EGS sector in APEC and to provide a coherent setting for relevant work under way in various APEC bodies. The Framework has four components: (a) research and development, (b) supply, (c) trade and (d) demand for EGS.	AELM 2011: In 2012, economies will work to develop an APEC list of environmental goods that directly and positively contribute to our green growth and sustainable development objectives.

Energy Smart Communities Initiative (2010)	Includes pillars for smart power grids, smart buildings, smart transport and smart jobs and education. Smart transport pillar includes projects on electromobility roadmaps and electric drive vehicle demonstrations.	TPTMM-EMM 2011: Ministers direct the EWG and TPTWG to identify and study appropriate strategies, approaches and best practices for promoting efficient and alternative-fueled vehicles, including electric drive vehicles, based on life cycle assessments.
Low Carbon Model Town project (2010)	Refers to towns, cities and villages which seek to become low carbon with a quantitative CO_2 emissions reduction target and a concrete low carbon developing plan.	EWG 2012 project on Establishing Low Carbon Energy Indicators for Development of APEC Low Carbon Town: To monitor the progress toward the targets and evaluate the effect of low carbon energy strategy and low carbon measures, proper and quantitative tool shall be selected.

Notes: TPTMM-EMM – APEC Ministerial Conference on Transportation and Energy, EWG – APEC Energy Working Group, AELM – APEC Economic Leaders Meeting, TPTWG – APEC Transportation Working Group.

APEC Environmental Goods and Services (EGS) Program and the Biofuels Initiative

APEC members adopted an EGS Program Framework in 2008 with an intention to support the development of the EGS sector. Since 2012, APEC has been working on identifying EGS that directly and positively contribute to the green growth and sustainable development objectives. As specified in the supply pillar of the EGS program, APEC members tend to be interested in goods that incorporate cleaner, more resource and energy efficient technologies and services that include the training of skilled personnel. From a product-level perspective, APEC so far seems to favour energy saving products, such as light-emitting diodes for street and indoor lighting, energy saving windows, but also low emission products, including electric vehicles and biofuels.[2]

At a product or industry level, if a practitioner is interested in quantifying the total direct and indirect environmental impacts throughout the whole production history, then Input-Output (hybrid) LCA models are a unique analytical tool. Since the life cycle of a product now spans across multiple national borders, these models are often sourced with MRIO data. MRIO LCA is capable of showing, for example, that products incorporating cleaner technologies or intended for cleaner use are still associated with a considerable amount of emissions that occurred at

---

[2] See Appendix 12: APEC Environmental Goods and Services Work Program, 2011 CTI Report to Ministers. Available at: http://publications.apec.org/file-download.php?filename=Appendix 11_EGS WP Mapping Matrix.pdf&id=1209_toc.

certain points in the production history when emission-intensive materials were used, such as steel, plastics or aluminum.

Experimental hybrid LCA studies benefitting from new MRIO datasets have helped to unravel the full supply chain of emissions required to produce certain environmental products. According to a 2011 report by Thomas Wiedmann and colleagues [3] on wind power in the UK, process-based accounting (limited boundary) showed that wind power turbines would cause 13.4 g of CO_2 life cycle emissions per 1 kWh of generated electricity, while alternative MRIO-based accounting (economy-wide) yielded estimates of 28.7 - 29.7 g.

APEC did not undertake comprehensive assessment of EGS in terms of environmental impact at product level, but the members are familiar with some basic insights from LCA studies. Biofuels may serve as an example.

According to the 2010 Fukui Declaration, biofuels from sustainable biomass sources are considered to have a far smaller carbon footprint than oil and can therefore displace a share of oil use for transportation. As such, an APEC Biofuels Task Force was created to look into biofuel economics and trade, biofuel vehicles and infrastructure, and biofuel resources.

The Biofuels Task Force has found that second-generation biofuels from farm and forest residues could potentially displace two-fifths of gasoline use and one-fifth of crude oil imports in the APEC region while generating substantial employment opportunities. First generation resources from conventional food crops have far less potential, only displacing under certain scenario about 5 percent of crude oil imports in the region.

It is true that biofuels directly emit less carbon and pollutants, but the life cycle greenhouse gas emissions of a biofuel comprise those from cultivation, harvest, transport and processing as well as other emissions saved from utilization of co-products. While theoretically biofuels are renewable, their potential for reducing life cycle greenhouse gas emissions varies significantly such that in assessing the environmental benefits, proper analysis should be done on an individual basis. Environmentally extended MRIO will be a useful tool for looking into how the biofuels sector development will affect other important sectors such as transportation, energy and agriculture.

To better inform trade and environmental policies, it may be desirable to augment conventional measures of the CO_2 reduction potential with estimates of total emissions resulting from the upstream and downstream supply chains of EGS. If Hybrid LCA is applied in a consistent manner to an array of potential EGS, policy makers will be informed which goods or services they may prioritize for trade liberalization and facilitation to attain the maximum emission reduction effect. Now, with the advent of experimental Hybrid MRIO LCA frameworks and availability of new MRIO databases with more product-level detail, such analysis is becoming more plausible. Given the political sensitivity of identifying EGS, MRIO as a neutral, objective and scientific tool may provide some useful

[3] Wiedmann, T.O., Suh, S., Feng, K., Lenzen, M., Acquaye, A., Scott, K. and Barrett, J.R. (2011) Application of Hybrid Life Cycle Approaches to Emerging Energy Technologies – The Case of Wind Power in the UK. Environmental Science & Technology, 45 (13): 5900-5907.

evidence on emissions associated with particular products, for example electric vehicles.

APEC Low Carbon Model Town (LCMT) Project and Low Emission Development Strategies

To promote the Green Growth agenda, in the 2011 Honolulu Declaration APEC members committed to incorporate low-emissions development strategies into their economic growth plans, including through creating energy-smart, low-carbon communities – the Low-Carbon Model Town and similar projects.

While several sustainable urban development projects are ongoing in the APEC region, APEC members decided to collectively demonstrate leadership in the creation of low-carbon communities in urban development plans. The APEC Low-Carbon Town refers to towns, cities and villages which seek to become low carbon with a quantitative CO_2 emissions reduction target and a concrete low carbon development plan irrespective of its size characteristics and the type of development. The multi-year APEC LCMT project consists of three activities, namely, (1) development of the Concept of the Low-Carbon Town, (2) feasibility studies and (3) policy reviews of planned town and city development projects. As the key advisory body for the APEC LCMT project, an LCMT Task Force was established in response to the APEC Energy Minister's instructions in their 2010 Fukui Declaration. The concept of LCMT stresses the importance of setting and evaluating quantitative low carbon reduction targets with a given time frame.

APEC members agreed to nominate their urban development areas to be part of the LCMT project. The Yujiapu Financial District in Tianjin, China has been selected as Phase 1 of the APEC LCMT for a feasibility study. APEC's first low carbon town is also expected to become the world's first low carbon central business district. It is targeted for completion in 2020 with an area under development of 3.65 km^2 and planned (daytime) population of 500,000.

Samui Island, Thailand was selected as the second LCMT and the first APEC Low Carbon Island with an area of 227 km^2 and population of 50,000 plus an estimated 100,000 immigrant residents and tourist inflow of 1,000,000 per year.

The feasibility studies set the following mid-term (by 2020) and long-term (by 2030) reduction targets for on-site emissions: 30 and 50 percent, respectively, for Yujiapu District; and 20 and 40 percent, respectively, for Samui Island. Lower emissions will be achieved through various measures to reduce energy consumption in the sectors of commercial and residential buildings and transportation.

However, evaluation of the low carbon development is not straightforward and may depend on the accounting strategy. In a low carbon community, emissions resulting directly from economic activities and lifestyle may be reduced to a minimum. We may even imagine a low carbon town with virtually zero emissions produced within the town itself. But does this mean that it is not responsible for emissions embodied in the products it consumes?

If a low carbon town project is successful, it will attract more residents and visitors that will lead to increased consumption of goods and services. The massive demand that city dwellers create will be met by a supply of products from other areas. Is it possible to identify and gauge all these indirect impacts of a low carbon town development?

MRIO tools may be an effective solution provided that the required data are available at a sub-national (district, province, state) level where the low carbon town or island in question is one of the regions in the Input-Output system. Manfred Lenzen and Glen Peters[4] show how MRIO modelling can be applied to enumerate the impacts of household consumption in Australia's two main cities – Sydney and Melbourne – in terms of environmental, economic and social indicators. The results demonstrate that consumption by a household in a particular city has emissions, water, monetary, and employment consequences that spread across the entire country. Impacts are both direct and indirect, with indirect supply-chain effects exceeding direct effects for all indicators except employment. In other words, families cause significantly more greenhouse gas emissions, use more water, and cause more money to circulate around the wider economy through their consumption of goods and services than directly by driving cars, burning household gas, running their water taps, or spending money at the shops. While direct annual greenhouse gas emissions per household in Sydney and Melbourne are 9.3 and 13.0 ton respectively, indirect emissions amount to 68.2 and 78.3 ton. This result highlights the importance of indirect effects for decision-making.

Besides, the more the town focuses on green or clean economic activities (mainly services), the more traditional emitting/polluting activities are relocated outside of the town and the larger will be the difference between the two accounting strategies, direct vs. total emissions. Low carbon communities are thought to directly produce low emissions as the residents are mostly engaged in services. But as explained by Sangwon Suh[5], services are responsible for large indirect emissions. In the case of the US, service industries produce less than 5 percent of the total national direct greenhouse gas emissions but consumption of services is responsible for about 37 percent of total emissions.

Estimation of a LCMT carbon footprint may therefore reveal whether transition to a low carbon community indeed leads to reduced environmental pressures on the national economy. As a result, greening of the whole supply chain beyond the boundaries of the low carbon community may be prioritized. APEC members are set to discuss the development of a set of low carbon energy indicators for LCMT and carbon footprinting may be useful for a holistic assessment of the LCMT performance, provided that the required data are available.

Similar conclusions may be applicable to another part of the APEC Green Growth agenda: Low Emission Development Strategies at the national and international level. Intensive trade flows within the APEC region also transmit a significant amount of emissions within APEC and between APEC and external

[4] Lenzen, M., Peters, G.M. (2010) How City Dwellers Affect Their Resource Hinterland: A Spatial Impact Study of Australian Households. Journal of Industrial Ecology, 14: 73-90.
[5] Suh, S. (2010) Comprehensive Environmental Data Archive (CEDA), in The Sustainability Practitioner's Guide to Input-Output Analysis, ed. by Joy Murrey and Richard Wood. Common Ground Publishing, Champaign, Illinois: 121-122.

partners. According to a recent OECD (2009) report, Some APEC members are among the biggest net CO2 importers (such as US, Japan) and exporters (for example China, Russia)[6]. In developing economies, Low Emission Development Strategies may not perform to their full potential unless developed members realize the shared responsibilities for a part of emissions embodied in imported goods and services that they consume.

APEC has been discussing low carbon targets for member economies. Understanding of the level of greenhouse gas emissions and absorptions within each economy will be vital in evaluating as well as enacting methods to achieve the targets. MRIO analysis has a role in understanding the issue in a holistic manner considering the diverse capacity and level of economic development of APEC member economies.

Expectations from Future Environmentally Extended MRIO Oriented Research

In November 2011, experts in Input-Output economics, during the APEC Conference on "Building APEC Economies' Capacities of Employing Input-Output Tables for Advanced Economic Modeling", provided suggestions on various applications of MRIO tools relevant to the APEC agenda, including the analysis of global supply chains and design of global environmental policies.

The APEC Senior Officials' *Steering* Committee on Economic and Technical Cooperation (SCE) also discussed ways to increase exposure of policy makers to the application of Input-Output based accounting and modelling tools through tailored capacity building activities at the two SCE meetings throughout 2012 in Russia.

To provide a deeper understanding of supply chain issues, the APEC Committee on Trade and Investment also held a Trade Policy Dialogue in Kazan, Russia in May 2012 to further discuss fragmented and globally linked production chains and the increasing importance of trade in value-added where MRIO could be used as a measurement tool.

Lack of reliable and comprehensive data has long been a significant obstacle to using MRIO analysis for developing informed environmental policies. However, with the recent availability of global environmentally extended MRIO datasets (Eora, EXIOPOL, WIOD and other), analysis of various issues relevant to the APEC agenda is becoming more accessible.

As explained above, the APEC Green Growth agenda encompasses various initiatives, including energy security, EGS, low-carbon demonstration vehicles, and the Low Emissions Development Strategy. The variable level of aggregation in MRIO models is useful to evaluate each of these initiatives at a detailed product or industry level but also to capture their overall impact.

---

[6] Nakano, S. et al. (2009), "The Measurement of CO2 Embodiments in International Trade: Evidence from the Harmonised Input-Output and Bilateral Trade Database", OECD Science, Technology and Industry Working Papers, 2009/03, OECD Publishing. http://dx.doi.org/10.1787/227026518048

In the case of LCA at a product level, existing studies have mostly focused on greenhouse gas emissions induced along the upstream supply chain. In the APEC EGS context, it will be important to quantify the effect on the downstream supply chain as well, i.e. how a more efficient product will save energy and reduce emissions. A topical policy issue where the quantitative application of supply-driven or price MRIO models may inform APEC discussions is rationalizing fossil fuel subsidies and introducing environmental technologies.

At a more aggregate regional level, beyond focusing on the achievements of individual economies, APEC as a community of 21 economies will need to show progress on its Green Growth agenda. MRIO, given its multi-sector and multi-region analytical capability, will be a useful tool for such aggregate regional analysis. APEC could ultimately benefit from MRIO experts' advice on the best feasible measures to promote and evaluate its green growth initiatives.

Further readings

APEC Committee on Trade and Investment. (2009). APEC Environmental Goods and Services Work Program.
    http://www.apec.org/Groups/Committee-on-Trade-and-Investment/~/media/Files/Groups/IEG/App6_09_cti_rpt_EGS%20WP.ashx

APEC. (2007). Sydney APEC Leaders' Declaration on Climate Change, Energy Security and Clean Development. APEC.
    http://www.apec.org/Meeting-Papers/Leaders-Declarations/2007/2007_aelm/aelm_climatechange.aspx

APEC. (2010). The APEC Leaders' Growth Strategy. APEC.
    http://www.apec.org/Meeting-Papers/Leaders-Declarations/2010/2010_aelm/growth-strategy.aspx

APEC. (2011). The Honolulu Declaration - Toward a Seamless Regional Economy. Annex C - Trade and Investment in Environmental Goods and Services. APEC.
    http://www.apec.org/Meeting-Papers/Leaders-Declarations/2011/2011_aelm/2011_aelm_annexC.aspx

Kuriyama, C. (2012). A Snapshot of Current Trade Trends in Potential Environment Goods and Services. APEC PSU Policy Brief no. 3. APEC. APEC PSU .

Wiedmann, T., & Barrett, J. (2012). Environmental extensions and policy-relevant applications of MRIO databases. In S. Inomata, & B. Meng, Asian International Input-Output Series No 80. Tokyo: Institute of Developing Economies, Japan External Trade Organization.

Resources

http://www.biofuels.apec.org – APEC Biofuel Task Force
http://esci-ksp.org/?task=lcmt-low-carbon-model-town – APEC Energy Smart Communities Initiative Knowledge Sharing Platform / Low Carbon Model Towns

Acknowledgements

The authors would like to thank Dr Denis Hew, Director at the APEC Policy Support Unit for his inputs and suggestions.

Chapter 25: Environmental-Extended Input-Output Analysis in the System of Environmental-Economic Accounting (SEEA)

Julian Chow[1]

Introduction

The System of Environmental-Economic Accounting (SEEA) Central Framework is a multi-purpose, conceptual framework that describes the interactions between the economy and the environment, and the stocks, and changes in stocks, of environmental assets. It provides a structure to compare and contrast source data and allows the development of aggregates, indicators and trends across a broad spectrum of environmental and economic issues. It brings together information on water, minerals, energy, timber, fish, soil, land and ecosystems, pollution and waste, production, consumption and accumulation in a single measurement system. The concept and definitions that comprise the SEEA Central Framework are designed to be applicable across all countries. The SEEA Central Framework was adopted as an international statistical standard by the United Nations Statistical Commission at its 43rd Session in February 2012.

Also at the 43rd session, the Statistical Commission re-affirmed its endorsement of work to develop the SEEA Applications and Extensions, which is a document that highlights the potential of data compiled following the SEEA Central Framework to be applied to a range of policy and research questions and to be extended to integrate with data in other domains. SEEA Applications and Extensions describes monitoring and analytical approaches in which SEEA data

---

[1] The view expressed in this paper are those of the author and do not necessarily represent the views of the United Nations Statistical Division.

can be used to inform policy analysis. It provides a bridge between compilers and analysts and therefore is an important document in promoting and supporting the implementation of the SEEA Central Framework.

For the analyst of environmental-economic topics, the SEEA Applications and Extensions provides an insight into the benefits that may be gained from using a common, integrated framework for relevant SEEA data. Particular areas of analysis that are highlighted include analysis of sustainable resource use and environmental efficiency, analysis of environmental activities, analysis of environmental assets and natural resources, and analysis of the household sector's behaviour with respect to the environment. The information described here may be relevant for both the development of policy and the monitoring and evaluation of policies.

Chapter 3 of the SEEA Applications and Extensions – "Analytical techniques" – considers the application of data from the perspective of the type of techniques that are relevant to both concept and dataset from the SEEA Central Framework and that may be applied across analysis of different topics. A particularly significant part of the chapter introduces the development of environmentally extended – input-output tables, EE-IOT. These tables provide a statistical base for a wide variety of analysis – both more straightforward structural analysis and more complex modelling.

The chapter also describes the environmentally-extended multi-regional input-output (EE-MRIO) analysis, which is a form of consumption-based modelling that has been widely applied in recent years to track environmental demands along the complete supply chain of products and services in an Input-Output framework. It highlights links between economic activities and their environmental consequence.

The purpose of this chapter is to describe EE-IOT and EE-MRIO analysis in the context of SEEA. The following section gives an introduction to the supply and use tables in the SEEA Central Framework that form the basis of the data compilation framework. Next I explain the underlying concept of the standard input-output analysis and its extension to environmental accounting. The following section extends the environmentally-extended input-output analysis into a multi-regional input-output framework. The pen-ultimate section outlines a number of techniques that may be applied to data from these SEEA-compliant EE-IOTs to answer policy questions. The conclusion identifies the ways in which the SEEA system has relevance for policy development and evaluation and decision making.

Introduction to the Supply and Use tables in the SEEA Central Framework Intro to first section here

At the heart of the SEEA Central Framework is a systems approach to the organization of environmental and economic information that covers, as completely as possible, the stocks and flows that are analysis of environmental and economic issues. In applying this approach, the SEEA Central Framework applies the accounting concepts, structures, rules and principles of the System of National Accounts (SNA). In practice, environmental-economic accounting includes the compilation of physical and monetary supply and use tables, functional accounts and asset accounts for natural resources.

Supply and Use Tables (SUT) are an integral part of the System of National Accounts (SNA). The supply table shows the supply of goods and services by product and industry, distinguishing between domestic industries and imports (hence it is a product-by-industry table). The use table shows the use of goods, services and value-added by product and by type of use, such as, intermediate consumption (industry) and final consumption (hence it is a product-by-industry table). The SUT are a central component of the SNA2008 as they show the flows of money through an economy and are used for both statistical and analytical purposes.

The SEEA Central Framework provides the conceptual foundation for environmental extensions to SNA-based SUT. The Central Framework records flows from the environment to the economy (natural input), within the economy (product flows), and flow from the economy to the environment (residuals).

Natural inputs are all physical inputs that are moved from their location in the environment as a part of the economic process or are directly used in production. Products are goods and services that result from a process of production in the economy. Residuals are flows of solid, liquid and gaseous materials, and energy that are discarded, discharged or emitted by the economy and households through processes of production, consumption and accumulation. Products are characterized by positive monetary value in transaction, whereas residuals are characterized with no transaction or zero monetary value in transaction. All these environmental flows can be recorded in both monetary and physical terms in SUT. Figure 1 provides a diagrammatical illustration of the physical flows of natural imports products and residuals between the environment and the economy.

Figure 1: Physical flows of natural inputs, products and residuals

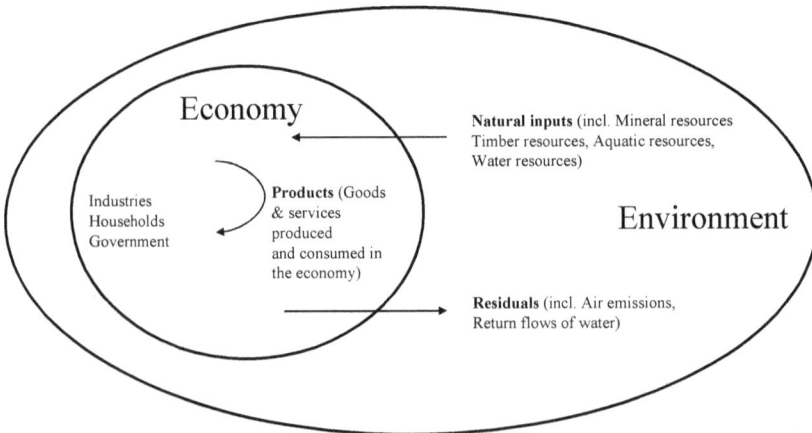

Monetary SUT record all flows of products in an economy between different economic units in monetary terms. They are compiled to describe the structure of an economy and the level of economic activity. Many of the flows of products

recorded in monetary terms relate to the use of natural inputs from the environment, for example the manufacture of wood products, or to activities and expenditures associated with the environment, for example environmental protection expenditure.

Table 1: Basic form of a monetary supply and use table

	Industries	Households	Government	Accumulation	Rest of the world	Total
Supply table						
Products	Output				Imports	Total supply
Use table						
Products	Intermediate consumption	Household final consumption expenditure	Government final consumption expenditure	Gross capital formation (incl. changes in inventories)	Exports	Total use
	Value added					

Note: Dark grey cells are null by definition.

Physical SUT (PSUT) record physical flows of products and residuals in physical units of measurement. PSUT are used to assess how an economy supplies and uses energy, water and materials and are also used to examine changes in production and consumption patterns over time. In combination with data from monetary SUT, changes in productivity and intensity in the use of natural inputs and the release of residuals can be examined.

Table 2: Basic form of a physical supply and use table

	Industries	Households	Accumulation	Rest of the world	Environment	Total
Supply table						
Natural inputs					Flows from the environment	Total supply of natural inputs
Products	Output			Imports		Total supply of products
Residuals	Residuals generated by industry	Residuals generated by final household consumption	Residuals from scrapping and demolition of produced assets			Total supply of residuals
Use table						
Natural inputs	Extraction of natural inputs					Total use of natural inputs
Products	Intermediate consumption	Household final consumption	Gross capital formation	Exports		Total use of products
Residuals	Collection & treatment of waste and other residuals		Accumulation of waste in controlled landfill sites		Residual flows direct to environment	Total use of residuals

Note: Dark grey cells are null by definition. Blank cells may contain relevant flows.

The supply and use tables in the SEEA Central Framework provide a strong data compilation framework. They provide an important information base for the various forms of Environmentally-Extended Input-Output analysis. Table 3 and Table 4 give examples of the air emissions and energy data compiled using the supply and use table in the SEEA Central Framework.

Table 3: Combined supply and use table for air emissions

	Industries (by BIC categories)								Households	Government	Total
	1-3	6-9	10-33	35	36-39	41-43	49-56	45-47,58-99			
1. Output by industry (currency units)	170 737	116 473	1 581 433	195 769	76 916	526 526	696 332	3 759 357			7 123 543
2. Intermediate consumption and final use (currency units)	146 006	103 131	1 521 247	180 772	62 482	511 084	616 833	3 162 097	491 935	163 978	6 959 565
3. Gross Value Added (currency units)	24 731	13 342	60 186	14 997	14 434	15 442	79 499	597 260			819 891
4. Employment	371	185	1 865	61	105	668	1 001	6 675			10 931
5. Environmental protection expenditure (currency units)											
Protection of ambient air and climate	175	58	351	585			370		554	419	2 512
6. Environmental taxes (currency units)											
Carbon taxes	343	22	1 108	23	146	142	1 243	2 588	6 985		12 600
7. Generation of air emissions (tonnes)											
Carbon dioxide	10 610	2 121	41 434	53 197	9 436	2 299	29 517	17 093	38 412		204 120
Methane	492	36	16	4	233	2	2	5	20		806
Dinitrogen oxide	24		4	1	2		1		1		32
Nitrous oxides	69	6	38	23	5	15	261	45	51		514
Hydroflourocarbons	3		28	6			62	1	1		103
Non-methane volatile organic compounds	5	8	40		1	8	17	17	67		163
Particulates (incl PM10, dust)	7		9			2	9	2	9		39
8. Air emissions from transport activity (tonnes)											
Carbon dioxide	2 673	54	1 065	14	77	1 843	27 748	7 297	18 921		59 692
Methane							1		2		3
Dinitrogen oxide							1		1		2
Nitrous oxides	28		5			15	260	36	38		380
Hydroflourocarbons	3						62	1			67
Non-methane volatile organic compounds	4					2	8	4	35		52
Particulates (incl PM10, dust)	1		1			1	9	2	6		19

Note: Dark grey cells indicate zero entries by definition.

Environmental-Extended Input-Output Table (EE-IOT)

In order to apply the analytical techniques environmental data are often combined with input-output tables (IOT). Input-Output (IO) analysis aims to understand the interaction between economic sectors, producers and consumers. IOT display the interconnection between different sectors of production and allow for a tracing of production and consumption in an economy. The compilation of IOT is described as an analytical extension in the System of National Accounts being derived through the combination of supply and use tables (SUT) which are core accounts of the SNA. Various mathematical and analytical approaches are available to convert SUT to an IOT.

Table 5 shows a simplified version of a Single Region Input-Output (SRIO) table. It gives a detailed description of domestic production processes and transactions within a single country (or region). An IOT is usually structured as a product-by-product or industry-by-industry table. Table 5 shows an industry by industry table of j industries. The rows show the outputs of an industry while the columns provide information about the inputs required in the production process of an industry.

The output of the industries is the sum of intermediate consumption (Zd) (which is a j by j matrix) and final demand categories such as final consumption (cd), gross capital formation (fd) and exports (ed). The subscript d denotes the use of domestic inputs, since the imported goods and services are shown in the row "imports" (and are denoted by subscript m). On the input side of an industry there are the intermediate inputs (Zd), value added categories (such as compensation of employees (wages) and operating surplus (profit)), and imports (m). Of course the inputs in an industry must equal the output, which is why the row and column sums (q) must be equal. All the variables with the subscript tot are scalars that show the totals for those respective row or columns.

Table 5: A single region input-output table (SRIO) with environmental data

		Industries			Final demand			Total
		1	...	*j*	Final consumption	Gross capital formation	Exports	
Industries	*1*							
	...		Z_d		c_d	f_d	e_d	Q
	j							
Value added		v						
Imports		m			c_m	f_m	e_m	m_{tot}
Total		q			c_{tot}	f_{tot}	e_{tot}	
Natural inputs/residuals		r						r_{tot}

The intermediate input matrix (Z) of an IOT is therefore square (it contains the same number of rows and columns) and symmetric (the items indicated by the rows and columns are the same: both are products or both are industries). The abbreviation SIOT is sometimes used to refer to a square and/or symmetric IOT

Table 4: Combined supply and use table for energy data

	Industries (by ISIC categories)							Rest of the world	Taxes less subsidies on products, trade and transport margins	Final consumption			Total
	ISIC 01	ISIC 02	ISIC 03	ISIC 04	ISIC 08	Other industries	Total industry			Households	Government	Capital Formation	
1. Supply of energy products (Currency units)													
Coal								26 125				1	26 126
Peat and peat products													
Oil shale / oil sands													
Natural gas	4 614			4 312			8 926		3 891				12 817
Oil	12 589	6 164					18 753	17 232	562				36 547
Biofuels	2		2	12			16						16
Waste	111		156				267	9					276
Electricity				14 414			14 414	9	8 113				22 536
Heat				665			665						665
Nuclear fuels and other fuel nec													
2. Total supply of products (Currency units)	59 780	72 669	38 288	39 765	304 401	6 608 640	7 123 543						
3. Intermediate consumption and final use (Currency units)													
Energy products	10 081	24 519	20 512	8 726	14 293	256 077	334 207	273 170		63 362	2 150	-5 200	667 688
Total (energy and non-energy products)	51 121	62 143	32 742	18 358	269 338	5 809 950	6 303 652			491 935	163 978		819 891
4. Gross value added (Currency units)	8 659	10 526	5 546	21 407	35 063	738 690	819 891						819 891
5. Depletion of natural energy resources (currency units)													
Depletion adjusted value added	8 659	10 036	5 546	21 407	35 063	738 690	819 401						819 401
6. Employment	145	148	78	165	374	9 921	10 831						10 831
7. Supply of energy products (PJ)													
Coal													
Peat and peat products													
Oil shale / oil sands								225					225
Natural gas	395			369			764						764
Oil	721	347					1 068	930					1 998

Note: Dark grey cells indicate zero entries by definition.

Biofuels	5						7
Waste	39	55		17			110
Electricity		212		22			234
Heat		79					79
Nuclear fuels and other fuel nec							
8. I nd-use of energy products (PJ)							
Coal	2	17	2		20	2	1	−21	2
Peat and peat products							
Oil shale / oil sands							
Natural gas	2	39		12	53	201	26	2	282
Oil	34	326	621	49	1 032	441	102	−3	1 572
Biofuels	3	4	2	1	2		5		7
Waste	7	22	37	15	45		33		79
Electricity	2	11	50	19	105	1	29		234
Heat			2		35	100	44		79
9. Closing stocks of natural energy resources (currency units / PJ)							
Oil resources	82 000						82 000
Natural gas resources	76 000						76 000
Coal and peat resources	84 000						84 000
Uranium	2 000						2 000
10. Depletion of natural energy resources (PJ)	1 161						1 161
11. Gross fixed capital formation (currency units)							
For extraction of energy resources	26 510				26 510			26 510	26 510
For supply of energy products	520	4 230	10	1	4 750			4 750	4 750
12. Closing Stocks of fixed assets for extraction of energy resources (currency units)							
For extraction of mineral and energy resources	238 500	190 560			429 060				429 060
For capture of energy from renewable sources	1 430	1 430			1 430				1 430
For distribution of energy products	620	1 902	2 350	80 260	85 132				85 132

The input-output table shown in Table 5 is in monetary units. However, it is possible to record the output of an industry, i.e. its products, in physical terms as well. For example, many studies have analysed energy using an IOT in which the output of the energy industries is measured in gigajoules or another energy unit. Table 6 shows such a hybrid unit IOT for which the industry j (shaded) is measured in physical terms. The input from this type of data could for example be from an energy account of the SEEA central framework.

The mathematical specifications of the input-output model work irrespective of the units of the rows of the input-output tables. The advantage of using physical units is that, in many cases, this provides a better description of technological relationships. The details of these types of models (for energy) are provided in Miller and Blair.

Table 6: A single region input-output table (SRIO) in hybrid units

		Industries			Final demand			Total
		1	...	j	Final consumption	Gross capital formation	Exports	
Industries	1							
	...	Z_d			c_d	f_d	e_d	q
	j							
Value added		V						
Imports		M			c_m	f_m	e_m	m_{tot}
Total		Q			c_{tot}	f_{tot}	e_{tot}	

The IOT can be augmented with environmental data by industry (denoted by the vector r in Table 5) which may be taken from SEEA accounts. In most applications these data relate to flows of natural inputs and residuals (see SEEA Central Framework Chapter 3). While not explicitly stated in the SEEA Central Framework, the conceptual foundation for environmental extensions to SNA-based IOT is documented in the System of Integrated Environmental and Economic Accounts (2003). The EE-IOT is an IOT extended to include environmental data. There is an input-matrix of environmental extensions and an output-matrix. Inputs are primary natural inputs. The output matrix of environmental extensions comprises the various forms of residuals such as emissions. The EE-IOT is portrayed in Table 7.

The sourcing and use of data on flows of natural inputs and residuals from SEEA accounts is advantageous for the compilation of EE-IOT since the information has already been organised in a manner consistent with the classifications and measurement boundaries that are used in the compilation of the IOT itself.

Table 7: Environmental extensions in an Input-Output framework (EE-IOT)

	Industries	Sub-total	Final demand			Total demand (basic prices)
			Final consumption	Gross capital formation	Exports, free on board price	
Industries	Industry by industry transactions in basic prices		By house-holds, non-profit institutions serving households (NPISH), Government	Gross fixed capital formation and changes in inventories	Intra- and extra EU	
Subtotal (1)	Total intermediate consumption by industry		Total final demand by type			Total demand
Tax less subsidies (2)	Net tax on production					
Total (1)+(2)	Total intermediate consumption in purchasers' prices					
Compensation of employees Other net taxes on production Consumption of fixed capital Operating surplus, net	Components of value added by industry					
Subtotal (3)	Value added					
Total (1)(2)(3)	Output by industry at basic prices					
Imports	Imports (cost, insurance and feight price)					
Total supply	Supply in basic prices					

Input (natural resources: land, fossil fuels, minerals, etc.)	Resource use per type and industry	Resource use per consumption activity			Total
Output (emission)	Emission per type and industry	Emission, per consumption activity			Total

Environmentally Extended Multi-Regional Input-Output (EE-MRIO) Analysis

Traditionally, Both IOT and EE-IOT are constructed for national economies under a single territory/region/country assumption. As production is increasingly global, the principles of producer and consumer responsibility for environmental pressures (such as pollution or resource use) are being increasingly discussed.

Consumption-based accounting (CBA) involves a re-allocation of environmental flows (e.g. air emission, waste, water etc.) from the point of production to the point of final demand, where all environmental residuals released along the supply chain of a product during its production and distribution are allocated to the end product as it is consumed. CBA complements the territory-based approach by including demand based driving force of environmental residual flows and provides an additional mechanism for countries to understand the impact that their consumption of goods and services has on the environment regardless of territorial boundaries.

EE-IOT is a form of CBA that has long been recognized as a useful top-down technique to attribute pollution or resource use to final demand in a consistent framework. In a single-territory EE-IOT, it is assumed that imported goods and services are being produced with the same technology as the domestic technology in the same sector, and therefore the model does not allow for a distinction between domestic and foreign production technology. To the extent that the domestic technologies are not representative of the technology of the imported product, the effect of the assumption is that the EE-IOT model will reflect the natural inputs and residual flows (e.g. emissions) avoided by importing rather than producing products domestically. In addition, trade in services (especially transport services) is omitted in the model.

In recent years, environmentally-extended multi-regional input-output (EE-MRIO) models have been developed to combine the environmental metrics with national economic accounts and trade statistics. The model takes into account full production chains with technologies specific to country of origin and opens the way for a new set of analyses. It shows the interconnections among different industrial sectors in various countries and offers a description of the global supply chains of products consumed. EE-MRIO analysis is increasingly used to analyze the environmental implications of consumption, be it for greenhouse gas emissions, land use or water uses. If the specific use of land, energy, water and emissions of the industry sectors in each country is known, the total land, carbon and water consumption associated with the production can be quantified. The environmental demands along the complete supply chain of products and services can be tracked by the model, thus highlighting links between economic activities and their environmental consequence.

There have been a number of large projects which have created multiregional input-MRIO tables. These MRIO databases create input-output tables for the whole world, although the actual regional breakdowns vary from around 40 to 190. Table 8 shows a simplified structure in which there is a country A and B. The accounting structure remains the same: the rows signify the output (both to the domestic and export markets) and the columns represent the inputs (also domestic and imported). In this way imports and exports are fully accounted for. The superscripts indicate the country of the variable. If there are two superscripts the first indicates the source and the second the destination. E.g. C_{AB} is the final consumption in country B of the output of country A.

The production of MRIO databases has enhanced the quality of the input-output models because it is no longer necessary to use the domestic technology assumption. In many cases the MRIO databases are linked to environmental and other socio-economic accounts, which makes it possible to analyse environmental and other sustainability issues.

Table 8: A multiregional input-output table (2 country) with environmental data

		Country A Industries	Country B Industries	Country A Final demand		Country B Final demand		Output
				Final consumption	Gross capital formation	Final consumption	Gross capital formation	
Country A	Industries	Z_{AA}	Z_{AB}	c_{AA}	f_{AA}	c_{AB}	f_{AB}	q_A
Country B	Industries	Z_{BA}	Z_{BB}	c_{BA}	f_{BA}	c_{BB}	f_{BB}	q_B
	Value added	v_A	V_B					
	Total input	q_A	q_B					
	Natural inputs/residuals	r_A	r_B					

An unavoidable consequence of the production of an MRIO is that it will not be consistent with the SRIO produced by national statistical offices. This is because SRIO are produced using data from that country only. On a national level, the supply and use accounting identity is used to balance the production and consumption statistics. In many cases the totals from source statistics (such as imports and exports) need to be adjusted to balance the system. In the MRIO balancing needs to be struck on a global scale. This also means that there may be conflicting source data which need to be confronted. The most important is the existence of "trade asymmetries" i.e. the phenomenon that the trade statistics on the imports to country A from country B is not equal to the data about the export of country B to country A. In the resolution of these asymmetries, as well as through other construction procedures, it is inevitable that differences between the MRIO and the SRIO of statistical institutes will emerge.

Analytical Techniques

This section discusses a number of analytical techniques based on these SEEA-compliant EE-IOTs to answer policy questions. EE-IOT are also used for quantifying separately the particular driving forces influencing changes in environmental impact but such decomposition analysis will not be explained further in this section.

Multiplier Analysis

The interpretation of the coefficients in the Leontief inverse matrix model is important. This matrix provides information about the direct and indirect effects of an increase in final demand. This is one of the most important advantages of the input-output model, since it makes explicit the linkages and feedback loops in an economy. Multipliers were traditionally compiled for economic variables such as output, value added, income, and employment, but the approach has been readily extended to environmental variables. The most commonly used environmental variables are energy and carbon dioxide. Other environmental variables include greenhouse gas emissions, acidification, and emissions of heavy metals to water. Overall, knowledge of the magnitude of a wide range of multiplier effects of individual industries provides relevant information for the evaluation of trade-offs.

There are several varieties of multipliers such as forward and backward linkages. The multipliers provide insight into the environmental pressures caused by the direct and indirect demand effects of a unit increase in final demand of a particular industry. Multipliers can therefore illustrate that an increase in the one industry will also lead to increases in environmental pressures in other industries through the (in)direct demand that is generated.

The relative importance of the Environmental Goods and Services Sector (EGSS) can be determined by calculating the indirect effects of its further growth. To obtain the appropriate multipliers for this sector, the data from the EGSS must be linked to data compiled in a national accounts based IOT. Using the re-worked and more detailed IOT, the multipliers and multiplier effects for the total EGSS can be calculated by multiplying the output of the different activities by their multipliers and dividing this total by the total output of the EGSS.

Attribution of Environmental Pressures to Final Demand

Input-output analysis is regularly used to attribute environmental pressures to final demand categories. This type of analysis can identify the link between final demand and resource use, emissions and other environmentally related flows. This input-output technique has been used to highlight areas that are subject for policy attention and answer a variety of questions. Three prominent ones will be discussed in more detail.

Footprint Calculations

The calculation of a "footprint" is a technique in which environmental pressures are attributed to domestic demand. This line of work was popularized through the introduction of "ecological footprints" in the early 1990s. The ecological footprint calculates the amount of land and water (surface area) that is necessary in the production of a certain consumption bundle. However, it does not use input-output techniques for its calculation.

Over the last decade or so, various other footprints have been introduced, and increasingly they are being calculated using MRIO models. Examples include the carbon footprints, water footprint and ecosystems pressure footprints. Several national statistical offices such as the OECD, Eurostat, Statistics Canada, Statistics Netherlands and Statistics Sweden have also explored the calculations of footprints. Although the methodologies are currently quite different, there are efforts to harmonize their calculation.

Production versus Consumption Perspective

Footprint indicators make explicit the environmental pressures that are caused by consumer behaviour. However, their calculation is often used to prove another point as well. It is a strand of the literature which is often referred to as the "production versus consumption perspective".

Underlying the discussion are the questions: Which environmental pressures is a country responsible for? In the polluter-pays-principle who is the polluter? On the one hand, it may be that industries, which are also responsible for most of the wealth creation in a country, are responsible. This is commonly referred to as the production perspective. Some international agreements, such as the Kyoto Protocol, follow this logic because they are based on all greenhouse gas emissions within the geographical boundaries of the country. On the other hand, the consumption perspective is based on the premise that the proverbial "polluter" is the person that consumes the end product of industry. The consumer perspective is captured by calculating environmental footprints because they include all environmental pressures abroad and in the source country.

Figure 2 gives an example of such analysis for the EU27, showing the CO2 emissions per capita in 2006 from both a consumption and production perspective. At a more detailed level, it appears that energy use in and around the house, personal mobility and food consumption drive in total around 70% of the CO2 emissions from households. Such insights are important in understanding which product-related and consumption-related policies may help to limit carbon emissions.

Figure 2: Production- and consumption based CO2 emissions per capita in the EU27 in

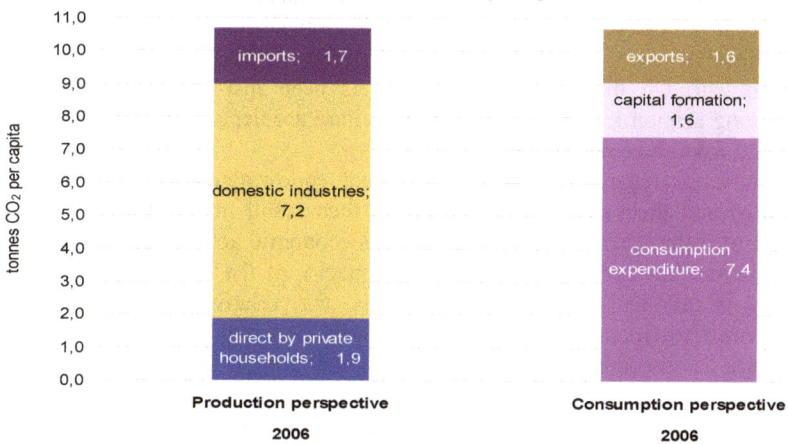

Global Shifts in Environmental Pressures

In a closed economy, the total environmental pressures following the producer or consumer perspective would be the same. Differences occur because of trading relationship with other countries in the world. One can therefore observe that all countries have an "environmental trade balance". This environmental trade balance, which is the difference between the environmental pressures embodied in imports and exports, will change over time. This may be caused by economic developments as well as institutional agreements such as the Kyoto Protocol on climate change or agreements on trade liberalisation.

A lot of research has been done to analyse these shifts in environmental pressures. Various hypotheses have been proposed. For example, the term "carbon leakage" is often used to label studies that investigate whether countries' emissions under the Kyoto Protocol are being reduced by importing emission intensive products from countries that do not participate in the Protocol. A related field of research is the "pollution haven hypothesis" that investigates the same shifts from developed to developing countries resulting from differences in environmental regulation.

It is possible to undertake consumption based modelling using a Life Cycle Analysis (LCA) approach whereby the "life cycles" for particular consumption items are traced and then the links to the use of various materials or emissions are drawn following the basic logic outlined above. The difference in using an LCA approach is that the fully integrated industry and product information inherent in an IOT is not utilised, although LCA may use some I-O relationships as part of their calculations.

Conclusion

The effect of human activity on the environment has emerged as one of the most significant policy issues. On the one hand, there has been growing concern about the effect of each country's economic activity upon the global and local environment. On the other hand, there has been increasing recognition that continuing economic growth and human welfare are dependent upon the benefits obtained from the environment.

Questions regarding how environmental endowments are being used and allocated and their impact on society can then result in the development of a variety of different policies. For example, is economic activity generating a level of pollution that exceeds the absorptive capacity of the environment or affects human health and well-being? What is the relative importance of the Environmental Goods and Services Sector in the economy? Which environmental pressures is a country responsible for? What are driving forces influencing changes in environmental impact? What are the environmental pressures caused by the direct and indirect demand effects of a unit increase in final demand of a particular industry?

The SEEA is a multi-purpose system and is relevant in a number of ways for policy development and evaluation as well as decision making. First, the summary information (provided in the form of aggregates and indicators) can be used to inform on issues and areas of the environment that are the focus of decision makers. Second, the detailed information, which covers some of the key drivers of change in the environment, can be used to provide a richer understanding of the policy issues. Third, data contained in the SEEA can be used in models and scenarios that are used to assess the national and international economic and environmental effects of different policy scenarios both within a country, between countries and at a global level.

EE-IOT is the type of technique that can be applied to the SEEA data to answer specific policy question. In recent years, there has been a surge in interest in consumption-based accounting to complement the traditional production-based IO analysis. The recent advance of the MRIO analysis not only allows the tracing of environmental demands along the complete supply chain of products and services but also the allocation of environmental impacts to each final demand category.

Going forward, the implementation of SEEA Central Framework at the country level will enrich the availability of the SEEA data and support a more detailed EE-IOT and EE-MRIO analysis. The benefits of these SEEA-compliant single and multi-regional EE-IOT to policy and decision making processes can be seen in specific areas such as energy and water resource management; patterns of consumption and production and their effect on the environment; and the so called "green economy" and economic activity related to adoption of environmental policies. The benefits are most broadly captured in relation to policies concerning sustainable development – one of the most pressing policy issues for current and future generations.

Further reading

Hertwich, E., & Peters, G. Multiregional Input-Output database Technical Document.

United Nations . (2003). Integrated Environmental and Economic Accounts 2003.

United Nations. (2012). Consultation Draft – System of Environmental Economic Accounting (SEEA) Applications and Extensions.

United Nations. (1999). Handbook of Input-Output Table Compilation and Analysis .

United Nations. (2012). Status report on the preparation of SEEA Extensions and Applications (2012), Draft materials prepared for the 7th Meeting of the Committee of Experts on Environmental-Economic Accounting (UNCEEA).

United Nations. (2012). System of Environmental Economic Accounting (SEEA) Central Framework.

United Nations. (2008). System of National Accounts 2008.

Wiedmann, T. (2009). A review of recent multi-region input–output models used for consumption-based emission and resource accounting. Ecological Economics , 69 (1), 211-222.

Part VI: The Future

Visitors at my window
Oil on canvas 84cm x 92cm
Dagmar Hoffmann

Chapter 26: Current and Future Policy Applications of MRIO Research

Thomas Wiedmann and John Barrett

Introduction

Recent years have seen the development of several environmentally extended multi-region input-output databases (EE-MRIO) and many applications have been described in the scientific literature[1-4]. However, it is not always obvious whether the insights gained from these studies have indeed been used in political decision-making. Neither is it clear what policy applications will look like in the future.

In this chapter we ask what kind of unique information can be produced by EE-MRIO models that cannot be generated by other approaches and whether there is evidence that this information has been used outside of the academic realm. In particular, we pose the question whether EE-MRIO analysis has been used to inform governmental policies. By so doing, we identify the additional benefits of EE-MRIO models lacked by other approaches in terms of informing policy analysis.

---

[1] Minx, J. C., Wiedmann, T., Wood, R., Peters, G. P., Lenzen, M., Owen, A., Scott, K., Barrett, J., Hubacek, K., Baiocchi, G., Paul, A., Dawkins, E., Briggs, J., Guan, D., Suh, S. and Ackerman, F. (2009) Input-output analysis and carbon footprinting: An overview of applications. Economic Systems Research, 21(3), 187-216. http://dx.doi.org/10.1080/09535310903541298.

[2] Peters, G. P., Andrew, R. and Lennox, J. (2011a) Constructing an Environmentally-Extended Multi-Regional Input–Output Table using the GTAP Database. Economic Systems Research, 23(2), 131-152. http://dx.doi.org/10.1080/09535314.2011.563234.

[3] Wiedmann, T., Lenzen, M., Turner, K. and Barrett, J. (2007) Examining the Global Environmental Impact of Regional Consumption Activities - Part 2: Review of input-output models for the assessment of environmental impacts embodied in trade. Ecological Economics, 61(1), 15-26. http://dx.doi.org/10.1016/j.ecolecon.2006.12.003.

[4] Wiedmann, T. (2009) A review of recent multi-region input-output models used for consumption-based emission and resource accounting. Ecological Economics, 69(2), 211-222. http://dx.doi.org/10.1016/j.ecolecon.2009.08.026.

Some specific applications are emerging as frontrunners, especially in the context of greenhouse gas (GHG) emissions embodied in trade. Accounting for emissions from a consumption perspective (i.e. adding emissions associated with imports to national GHG accounts and subtracting emissions associated with exports) is a highly relevant topic in climate policy. This perspective is also referred to as the 'national carbon footprint'. An international expert workshop on consumption-based GHG accounting in London in 2011 examined the full implications of the different options for consumption-based approaches to climate change governance and policy, and explored strategies to address emissions attributable to consumers. EE-MRIO was confirmed as the most appropriate analysis tool for national carbon footprint accounting.

In the following sections we briefly recap recent developments before we describe in detail policy-making processes from the UK that have evidently used results from EE-MRIO analyses. We then reflect on possible future applications.

Recent Developments in EE-MRIO Modelling

Model Development

In the last few years, the development of EE-MRIO datasets with global coverage has come a long way.[5] There are around half a dozen MRIO initiatives worldwide, mostly organised by research institutions with an interest in environmental sustainability and/or international trade. Typically the databases consist of national input-output tables of countries and world regions, linked with tables describing the trade of goods and services and with environmental data such as GHG emissions, water consumption, land use, or resource extraction. Some models have only become available in the last twelve months and so the full potential of applications is only just emerging.

Policy-Relevant Publications

Recently, there have been a number of journal publications and reports that describe EE-MRIO analyses and their implications for policy. A whole series of papers have presented detailed global analyses of national carbon footprints and GHG emissions embodied in international trade, and several papers have presented a more refined and focused analysis of particular aspects of trade and global emissions[6-9]. Key results show, for example, that 37% of global CO_2

---

[5] Wiedmann, T., Wilting, H. C., Lenzen, M., Lutter, S. and Palm, V. (2011a) Quo Vadis MRIO? Methodological, data and institutional requirements for multi-region input-output analysis. Ecological Economics, 70(11), 1937-1945. http://dx.doi.org/10.1016/j.ecolecon.2011.06.014.
[6] Davis, S. J. and Caldeira, K. (2010) Consumption-based accounting of CO2 emissions. Proceedings of the National Academy of Sciences, 107(12), 5687-5692. http://dx.doi.org/10.1073/pnas.0906974107.

emissions are from fossil fuels traded internationally and an additional 23% of global CO_2 emissions are embodied in traded goods. These findings have implications for carbon pricing and regulating policies at the different points of extracting, burning and consuming fossil fuels in the global supply chain.

Further insight is gained by following production and supply chains from raw materials to processed materials and further on to final consumer products and identifying and mapping associated emissions along the way. One example is a very detailed analysis of trade-embedded carbon flows published by the UK Carbon Trust, with a particular focus on the commodities aluminium, cars, clothing, cotton, and steel as well as the EU, UK and China[10]. Political relevance was demonstrated, for example, in the fact that the European Union is a large net importer of emissions embodied in aluminium and steel; yet these emissions are not priced under the EU emissions trading scheme.

Some journal publications, using EE-MRIO analysis, have focussed on the carbon footprint of individual countries such as Japan or the UK. Both countries are net importers of trade-embedded GHG emissions and therefore hold a 'carbon debt' towards other countries.

Increasingly, research is also being published that uses EE-MRIO modelling to examine the role of trade in virtual water and to establish water footprint accounts in the same way as is being done for GHG emissions (carbon footprints). EE-MRIO is well suited to complement process-based approaches to water footprinting by expanding the supply-chain coverage and by establishing the geography of embodied water.

The integration of carbon, water and ecological footprint indicators in one EE-MRIO framework has been accomplished in the project One Planet Economy Network Europe (OPEN:EU) funded by the European Commission[11]. Combining these overlapping, interacting and complementing indicators in a 'Footprint Family' offers additional benefits for decision-making[12,13]. The analysis and

---

[7] Davis, S. J., Peters, G. P. and Caldeira, K. (2011) The supply chain of CO2 emissions. Proceedings of the National Academy of Sciences, 108(45), 18554-18559. http://dx.doi.org/10.1073/pnas.1107409108.

[8] Hertwich, E. G. and Peters, G. P. (2009) Carbon Footprint of Nations: A Global, Trade-Linked Analysis. Environmental Science & Technology, 43(16), 6414–6420. http://dx.doi.org/10.1021/es803496a.

[9] Peters, G. P., Minx, J. C., Weber, C. L. and Edenhofer, O. (2011) Growth in emission transfers via international trade from 1990 to 2008. Proceedings of the National Academy of Sciences, 108(21), 8903-8908. http://dx.doi.org/10.1073/pnas.1006388108.

[10] Carbon Trust (2011) International Carbon Flows - Global flows. Report Number CTC795, 11 May 2011. The Carbon Trust, London, UK. http://www.carbontrust.co.uk/publications/pages/home.aspx.

[11] EUREAPA and OPEN EU Project; http://www.oneplaneteconomynetwork.net/eureapa.html

[12] Galli, A., Wiedmann, T., Ercin, E., Knoblauch, D., Ewing, B. and Giljum, S., 2012. Integrating Ecological, Carbon and Water footprint into a "Footprint Family" of indicators: Definition and role in tracking human pressure on the planet. Ecological Indicators 16, 100-112. http://dx.doi.org/10.1016/j.ecolind.2011.06.017.

[13] Ewing, B.R., Hawkins, T.R., Wiedmann, T.O., Galli, A., Ertug Ercin, A., Weinzettel, J. and Steen-Olsen, K., 2012. Integrating ecological and water footprint accounting in a

scenario tool EUREAPA was developed from the EE-MRIO model which allows users to analyse global supply chains using a carbon, ecological and water footprint indicator. The links between the consumption of a product type in one country and its production impacts elsewhere are identified and the top ten sources of greatest impact are displayed. The scenario editor can be used to explore the environmental pressures associated with changes in population, consumption patterns, production technology or trade over time.

Use of EE-MRIO Results in Policy-Making Processes

The main question posed in this chapter is whether EE-MRIO analysis has actually been used to inform governmental policies.

Consumption-Based Accounting of GHG Emissions

The strongest signals yet indicating that outcomes from EE-MRIO research have found their way into the policy-making process come from the United Kingdom. In contrast to results from accounting for territorial GHG emissions under the Kyoto Protocol, consumption-based accounting has shown that emissions attributable to the UK have increased during the last two decades [9, 14, 15, 16, 17].

The UK Department for Environment, Food and Rural Affairs (DEFRA) have committed to calculating the consumption-based accounts (CBA) of the UK for the next five years. This commitment means that CBA now form part of the DEFRA's headline indicators for sustainability[16].

In September 2011, the UK Energy and Climate Change Committee launched an inquiry to investigate the case for consumption-based GHG emissions accounting in the UK. The Committee examined the case for adopting consumption-based reporting in the UK, whether it would be feasible to do this in practice, whether emissions reduction targets might be adopted on a consumption basis, and what the implications for international negotiations on climate change

multi-regional input–output framework. Ecological Indicators 23, 1-8. http://dx.doi.org/10.1016/j.ecolind.2012.02.025.
[14] Baiocchi, G. and Minx, J. C. (2010) Understanding Changes in the UK's CO2 Emissions: A Global Perspective. Environmental Science & Technology, 44(4), 1177-1184. http://dx.doi.org/10.1021/es902662h.
[15] Barrett, J., Owen, A. and Sakai, M. (2011b) UK Consumption Emissions by Sector and Origin. Report to the UK Department for Environment, Food and Rural Affairs by University of Leeds. May 2011. Defra, London, UK.
http://randd.defra.gov.uk/Document.aspx?Document=FINALEV0466report.pdf.
[16] DEFRA (2010) Measuring progress - Sustainable development indicators 2010. UK Department for Environment, Food and Rural Affairs, London, UK.
http://sd.defra.gov.uk/documents/SDI2010_001.pdf,
http://sd.defra.gov.uk/progress/national/annual-review,
http://www.defra.gov.uk/statistics/environment/green-economy/scptb01-ems.
[17] Wiedmann, T., Wood, R., Minx, J., Lenzen, M., Guan, D. and Harris, R. (2010) A Carbon Footprint Time Series of the UK - Results from a Multi-Region Input-Output Model. Economic Systems Research, 22(1), 19-42.
http://dx.doi.org/10.1080/09535311003612591.

might be if the UK and others took this approach. In its response, the UK Energy Research Centre confirmed that it is possible to develop a robust methodology for accounting for GHG emissions on a consumption basis and that EE-MRIO analysis is the method of choice[18].

As well as EE-MRIO models being applied to the generation of policy-relevant indicators, there are increasing examples of applications related to policy formulation and appraisal. Before discussing these two examples it is important to consider the attribution of policy-orientated research directly to policy impact. It is very rare that an individual report or research output directly links to a measurable and isolated change in policy. Policy formulation is a complex process of negotiation with affected stakeholders, political persuasion, scientific evidence and other unmeasurable forces[18]. To suggest that one report directly changed policy would be naive. Additionally, the change could occur years after the study. This often happens with studies that have raised controversial issues, with consumption-based accounting being a good example.

This was very much the situation in the UK. In 2006, results were ready for publication, reporting the UK's consumption-based GHG accounts. It was not until 2011 that the UK Government adopted CBA as a headline indicator and committed to this for the following five years. However, this progress would not have happened were it not for the original study.

In 2011, researchers provided a yet more detailed analysis of the UK's carbon footprint, using an EE-MRIO model. The analysis provided the origin and destination of GHGs for the UK by 113 world regions and 57 sectors of the economy. For each product a matrix was constructed to show where emissions occurred and in which sector. The report was commissioned by DEFRA to provide insights into the origin of UK consumer emissions. There was not a specific policy question in mind when the study was commissioned but it was believed that the information would be useful to a number of issues in some of the government departments. It was felt within DEFRA and its Sustainable Consumption and Production (SCP) Programme that there was a pressing need for accurate and detailed CO_2 emissions data showing not only the sectors which emit the highest levels of carbon dioxide globally as a result of demand from the UK but also the specific countries where these emissions occur. Clearly, a truly multi-regional (multi-national) input-output approach was needed to answer these questions.

While the UK seems to be the only country yet to discuss the adoption of consumption-based GHG emissions accounting in official statistics, there are signs that other countries might also implement or at least support such an approach, e.g. China and some Scandinavian countries[19].

---

[18] Barrett, J., Le Quéré, C., Lenzen, M., Peters, G., Roelich, K. and Wiedmann, T. (2011) UK Energy Research Centre Response to the Energy and Climate Change Committee Consultation on Consumption-based Emission Reporting. 25 October 2011. UKERC, London, UK.
http://www.publications.parliament.uk/pa/cm201012/cmselect/cmenergy/writev/consumpt/con20.htm.
[19] Peters, G. and Solli, C. (2010) Global carbon footprints - Methods and import/export corrected results from the Nordic countries in global carbon footprint studies. Report

Results from EE-MRIO Modelling Help in Saving Resources

Another example where EE-MRIO results informed decision-making relates to the efficient use of resources. The Waste Resource Action Programme (WRAP) is an agency established by the UK Government that supports businesses and householders to reduce waste, develop sustainable products and use resources in an efficient way. WRAP commissioned a comprehensive study to assess how saving resources could help with reducing GHG emissions. The project, led by the Stockholm Environment Institute, employed a number of methods of which the most prominent was based on EE-MRIO (in the form of an environmentally extended two-region model comprising the UK and the Rest of World). This was combined with econometric forecasting models to provide an understanding of the changing production structure of the UK economy and to assess the role of material efficiency measures in reducing UK GHG emissions by 2050. The study provided robust evidence on the contribution of resource efficiency strategies to climate change targets[20]. The study demonstrated that strategies such as extending the lifetime of products, lean design techniques, reducing food waste and increasing product durability could all boost the service-based economy in the UK and reduce the shifting of GHG emissions abroad ('carbon leakage').

WRAP have since used the study extensively to help reshape their agenda, now increasingly concentrating on many of the resource efficiency measures included in the study. Additionally, the study has been circulated widely across UK Government departments, presented at cross-departmental committees and used as evidence in Parliamentary Select Committees. The study has also been used within DEFRA to help shape their sustainable products policy, clarifying what different resource efficiency strategies could deliver in terms of emission reductions. Finally, the study has also ensured that further evidence on specific measures is being investigated, in particular the shift from ownership of goods to services.

What Does the Future Hold?

What will EE-MRIO models be used for in the future? EE-MRIO has been recognised as the most suitable method to generate trade-adjusted accounts of GHG emissions and resource use. Those are accounts that add the impacts from imports and subtract the impacts from exports. This consumption-based accounting is likely to stay on the political agenda; at least in global climate policy due to announcement following the International Climate Change Conference in Durban in November 2011. As there will be no global comprehensive agreement until 2020 the issue of carbon leakage is still relevant.

---

number TemaNord 2010:592, 18 November 2010. Nordic Council of Ministers, Copenhagen, Denmark. http://www.norden.org/sv/publikationer/publikationer/2010-592.

[20] Barrett, J. and Scott, K. (2012) Link between climate change mitigation and resource efficiency: A UK case study. *Global Environmental Change*, 22(1), 299-307. http://dx.doi.org/10.1016/j.gloenvcha.2011.11.003.

As for the UK, it was envisaged that EE-MRIO data would allow the SCP Programme inside DEFRA to target its activities toward those sectors that emit the highest levels of carbon globally due to UK consumption, strengthen and focus dialogue with those countries where these emissions occur, and provide evidential input into the major internal and cross-Whitehall initiatives that are currently ongoing, such as the Green Economy Roadmap and the BIS-led Growth Reviews. In 2011/12 other important UK government departments such as the Department of Energy & Climate Change (DECC) and the Department for Business, Innovation and Skills (BIS) increasingly accepted that, whilst production-based CO_2 emissions are falling, consumption-based emissions are rising. They were eager for more detailed information about which sectors and countries are priorities to tackle, thus requesting the scientific community to provide such information to help maintain the momentum in delivering successful climate change policy.

Analyses undertaken as part of the EXIOPOL project produced results of arguably high political relevance, though the authors do not know whether the findings have actually influenced political decision-making in Europe. The top-down analysis of EXIOPOL is based on EE-MRIO analysis and confirms that Europe is generally a net importer of embodied natural resources and pollution[21]. Land use embodied in European imports is higher than the land use in Europe itself, and water consumption and material extraction embedded in European imports are also much higher than for exports. The policy relevance of results from EXIOPOL was further demonstrated with a dynamic analysis using a combination of EE-MRIO and a world trade model. GHG emissions and resource use reduction potentials were examined for three exemplary policy actions, namely the implementation of the Directive 2002/91/EC on the energy performance of buildings, the scrappage premiums for old cars in combination with tax incentives for low-emitting cars, and diet changes towards a Mediterranean diet with lower meat consumption.

It is to be expected that more policies like this will be evaluated with the EE-MRIO model. Eurostat used the tool developed within EXIOPOL to create national carbon footprint accounts. It is also intended to use the EE-MRIO from EXIOPOL to calculate true consumption-based Total Material Requirements (TMR) of the EU, thereby correcting a major drawback of indicators like Domestic Material Consumption (DMC), which only includes the weight of imported products without the raw material equivalents (the 'rucksacks' of primary materials) used in their production.

Other potential uses could be water management policies using water footprint analysis from EE-MRIO models[22, 23]. These could address issues such as

---

[21] EXIOPOL (2011) A new environmental accounting framework using externality data and input-output for policy analysis - Top-down approach. October 2011. Final report of the EXIOPOL project.
http://www.feem-project.net/exiopol/M54/EXIOPOL_Top_down_approach.pdf
[22] Daniels, P. L., Lenzen, M. and Kenway, S. J. (2011) The Ins and Outs of Water Use – A review of multi-region input–output analysis and water footprints for regional sustainability analysis and policy. Economic Systems Research, 23(4), 353-370. http://dx.doi.org/10.1080/09535314.2011.633500.

regional water management, the link between water use and energy use, or total water (footprint) intensity of sectors, processes, products or infrastructure.

Conclusion

Every model is a simplification of reality and therefore is limited in its ability to understand historical trends and future outcomes. Policy questions generally require flexible and versatile models and tools that adopt a comprehensive and holistic perspective whilst at the same time providing sufficient detail to allow for the evaluation of specific policies.

EE-MRIO analysis has clear strengths in areas where other modelling approaches have weaknesses and vice versa. The conclusions outline two key findings, firstly additional benefits of EE-MRIO models that other models lack in terms of informing policy analysis, and secondly a discussion of the value of EE-MRIO when combined with other modelling approaches, in particular econometric forecasting models for scenario construction.

In terms of additional benefits, EE-MRIO provides details of the fundamental link between production and consumption by taking a supply chain perspective. The unique feature of EE-MRIO is that it is the only approach that can provide a detailed and comprehensive picture of an economic sector and its interrelationship with other sectors across countries, thus capturing whole life-cycle impacts of products and services across international supply chains. Country and sector disaggregation is important for specific indicators such as biodiversity (which varies widely across countries and regions) and special trades, e.g. uranium exports.

In terms of scenario development, EE-MRIO provides an accounting system in which the modeller changes the model variables explicitly, not implicitly by a set of pre-defined rules as in econometric and computable general equilibrium models often used in policy making. The demand-driven nature of the input-output model makes it easy to make explicit changes to the level and pattern of household and government consumption expenditure.

One of the key advantages of EE-MRIO is its ability to be combined with other modelling approaches. This is partly due to standardised methods of accounting allowing integration of modelling approaches. Many of the policy examples in this chapter use EE-MRIO in combination with other methods, e.g. econometric forecasting models or process-based life-cycle analysis[20, 24].

For the further development of EE-MRIO a long-term strategy and commitment is needed from researchers, statistical offices and other stakeholders interested in decision-relevant outcomes. An important aspect of policy-relevant applications is the reliability and robustness of datasets. In the case of

---

[23] Duarte, R. and Yang, H. (2011) Input–Output and Water: Introduction to the Special Issue. Economic Systems Research, 23(4), 341-351.
http://dx.doi.org/10.1080/09535314.2011.638277.
[24] Wiedmann, T. O., Suh, S., Feng, K., Lenzen, M., Acquaye, A., Scott, K. and Barrett, J. R. (2011) Application of Hybrid Life Cycle Approaches to Emerging Energy Technologies – The Case of Wind Power in the UK. Environmental Science & Technology, 45(13), 5900-5907. http://dx.doi.org/10.1021/es2007287.

consumption-based GHG reporting, the establishment, in conjunction with the United Nations' Framework Convention on Climate Change and Eurostat, of new standards to harmonise reporting is recommended. If such standards were to be implemented in the future, it is likely that EE-MRIO would play a prominent role as the underlying calculation methodology.

Resources

Eora: A Global Multi-Region Input Output database; http://www.worldmrio.com, http://www.globalcarbonfootprint.com.

EUREAPA and OPEN EU Project;
http://www.oneplaneteconomynetwork.net/eureapa.html

EXIOPOL; http://www.feem-project.net/exiopol, http://creea.eu.

Global Trade Analysis Project (GTAP); https://www.gtap.agecon.purdue.edu.

World Input-Output Database; http://www.wiod.org.

List of Contributors

Angel H. Aguiar Center for Global Trade Analysis, Purdue University

John Barrett Sustainability Research Institute, School of Earth and Environment, University of Leeds, Leeds, UK; Centre for Sustainability Accounting, York, UK

Akhmad Bayhaqi APEC Policy Support Unit, Singapore

Joerg Beutel Konstanz University of Applied Sciences, Germany

John Boland Barbara Hardy Institute, School of Mathematics and Statistics, University of South Australia

Martin Bruckner Sustainable Europe Research Institute (SERI) Vienna, Austria

Leisa Burrell Sustainable Europe Research Institute (SERI) Vienna, Austria

Ignacio Cazcarro Department of Economics, School of Humanities and Social Sciences, Rensselaer Polytechnic Institute (RPI), USA. Collaborator with the Department of Economic Analysis, Faculty of Economics and Business, University of Zaragoza, Spain.

Julian Chow Department of Economic and Social Affairs, United Nations Statistics Division

Christopher Dey Integrated Sustainability Analysis (ISA) School of Physics, University of Sydney, Australia

Christophe Degain Economic Research and Statistics Division, World Trade Organisation

Erik Dietzenbacher Faculty of Economics and Business, University of Groningen, The Netherlands

Rosa Duarte Department of Economic Analysis, Faculty of Economics and Business, University of Zaragoza, Spain

Hubert Escaith Economic Research and Statistics Division, World Trade Organisation

Kuishuang Feng Department of Geographical Sciences, University of Maryland, USA

Stefan Giljum Sustainable Europe Research Institute (SERI) Vienna, Austria

Sandra Harrison Planning and Information Office, University of Sydney

Dagmar Hoffmann Artist, Bobs Farm, Port Stephens, Australia, dmhart@live.com.au and http://www.dagmarhoffmann.com.au

Klaus Hubacek Department of Geographical Sciences, University of Maryland, USA

Satoshi Inomata Institute of Developing Economies, JETRO

Shigemi Kagawa, Faculty of Economics, Kyushu University, Japan & Research Institute of Science for Safety and Sustainability, National Institute of Advanced Industrial Science and Technology, Japan

Keiichiro Kanemoto ISA, School of Physics, The University of Sydney, Australia; Graduate School of Environmental Studies, Tohoku University, Sendai, Japan

Yasushi Kondo, Faculty of Political Science and Economics, Waseda University, Japan

Jun Lan Integrated Sustainability Analysis (ISA) School of Physics, University of Sydney, Australia

Manfred Lenzen Integrated Sustainability Analysis (ISA) School of Physics, University of Sydney, Australia

Bart Los Faculty of Economics and Business, University of Groningen, The Netherlands

Christian Lutz Institute of Economic Structures Research (GWS) Osnabrück, Germany

Arunima Malik Integrated Sustainability Analysis (ISA) School of Physics, University of Sydney, Australia

Alexandra Marques IN+ Centre for Innovation, Technology and Policy Research, Mechanical Engineering Department, Instituto Superior Técnico, Universidade Técnica de Lisboa, Portugal

Andreas Maurer Economic Research and Statistics Division, World Trade Organisation

Bo Meng Institute of Developing Economies, JETRO

Dan Moran Integrated Sustainability Analysis (ISA) School of Physics, University of Sydney, Sydney, Australia

Kirill Muradov, International Institute for Training in Statistics, National Research University Higher School of Economics, Moscow, Russia

Joy Murray Integrated Sustainability Analysis (ISA) School of Physics, University of Sydney, Australia

Keisuke Nansai, Center for Material Cycles and Waste Management Research, National Institute for Environmental Studies, Japan & ISA, School of Physics, The University of Sydney, Australia

Badri Narayanan Center for Global Trade Analysis, Purdue University

Anne Owen Sustainability Research Institute, School of Earth and Environment, University of Leeds, UK

Isabelle Rémond-Tiedrez Eurostat, Luxembourg

Christian John Reynolds Barbara Hardy Institute, School of Mathematics and Statistics, University of South Australia; Integrated Sustainability Analysis, University of Sydney

José M. Rueda-Cantuche European Commission's Joint Research Centre and Pablo de Olavide University, Spain

Julio Sánchez-Chóliz Department of Economic Analysis, Faculty of Economics and Business, University of Zaragoza, Spain

Kjartan Steen-Olsen Industrial Ecology Program, NTNU, Trondheim, Norway

Sangwon Suh Bren School of Environmental Science and Management, University of California, USA

Marcel Timmer Faculty of Economics and Business, University of Groningen, The Netherlands

Susumu Tohno Graduate School of Energy Science, Kyoto University, Japan

Arnold Tukker Netherlands Organisation for Applied Scientific Research, Delft, Netherlands, and Norwegian University of Science and Technology, Institute of Product Design/Industrial Ecology program, Trondheim, Norway

Terrie Walmsley Center for Global Trade Analysis, Department of Agricultural Economics, Purdue University; and Department of Economics, University of Melbourne.

Colin Webb Economic Analysis and Statistics Division, Directorate for Science, Technology and Industry, OECD, France

Kirsten Wiebe Institute of Economic Structures Research (GWS) Osnabrück, Germany

Thomas Wiedmann School of Civil and Environmental Engineering, The University of New South Wales, Sydney, Australia; Integrated Sustainability Analysis (ISA), School of Physics, University of Sydney, Sydney, Australia; Centre for Sustainability Accounting, York, UK

Harry Wilting PBL Netherlands Environmental Agency, Bilthoven, The Netherlands

Richard Wood Industrial Ecology Program, Norwegian University of Science and Technology (IndEcol). NTNU Trondheim, Norway

Norihiko Yamano Economic Analysis and Statistics Division, Directorate for Science, Technology and Industry, OECD, France

Yang Yu Department of Geographical Sciences, University of Maryland, USA

Xin Zhou Green Growth and Green Economy Area, Institute for Global Environmental Strategies (IGES), Kanagawa, Japan

Index

Accounting Framework, 52, 220, 293

acidification, 56, 61, 134, 159, 161, 262

adjusted net savings, 231

affluence, 159, 170, 173, 177, 178

aggregation, 13, 15, 16, 25, 48, 50, 71, 74, 77, 108, 142, 227, 247

AIIO, 36, 41

APEC, xvii, 240, 241-248

basic prices, 11, 46, 49, 54, 223-229

bilateral trade data, 5, 13, 27, 80, 83

biofuel, 137, 157, 168, 242-244, 248

biomass, 85, 195-198, 244

Blue water, 115, 117, 119, 120

BRICSA countries, 191-199

carbon footprint, 15, 64, 65, 71, 73, 76, 77, 180, 183, 187, 188, 244, 246, 268, 269, 270, 271, 273

Carbon Trust, 269

C-intereg, 115

coefficient matrix, 80-83, 85, 86

Cohesion policy, 221, 234

Committee on Trade and Investment, 247

Commodity Technology Assumption, 58

competitiveness, 33, 52, 141, 206, 210, 215, 216, 218, 234

Composite commodity, 72, 75, 76

composite country, 72, 73-74

ComTrade, 13, 26, 47, 57, 210

construction minerals, 85, 195, 196, 198

consumer responsibility, xvii, 63, 113, 141-142, 146-155, 261, 191-192, 259

consumption patterns, 1, 92, 128, 135, 207, 253, 270

consumption-based accounting, 3, 85, 155, 191, 198, 260, 265, 268, 270-272

consumption-based CO$_2$ emissions, 92-93, 207

consumption-based emissions, 187, 273

consumption-based GHG accounting, 268

consumption-based inventory, xvii, 136-137, 139

convergence areas, 222

currency conversion, 11, 15-16

decision making, 172, 246, 272-273

DEFRA, 270-273

disaggregation, 16, 34, 48, 77, 106, 131, 161, 221, 237, 238, 274

Domestic Material Consumption (DMC), 273

domestic technology assumption, 16, 260

Dutch consumption, xvii, 125-126, 130-134

dynamic input-output model, 237, 238

Ecological Footprint, 57, 64, 104, 109-110, 262, 269

Economic and Technical Cooperation, 247

Economic Modeling, 247

EE-MRIO, 251, 259, 260, 265, 267-274

Effective protection rates, 217

embodied carbon, 2, 29, 85, 191-192, 198

embodied emissions, 90-93, 141, 150, 159, 161

embodied global-GHG intensity, 181-182, 184, 186-187

embodied materials, 191, 195, 197, 198

embodied water, 113, 118, 269

emissions accounting, 242, 270-271

emissions trade, 159

employment, xvii, 43, 49, 50, 106, 158-159, 171, 206, 216, 222, 234, 237, 238, 242, 244, 246, 262

End-of-pipe solutions, 168

energy consumption, 69, 77, 170-173, 177-178, 185-186, 245

energy security, 240, 241, 247

Environmental Accounting, 52, 191, 251, 273

Environmental extensions, 52, 54, 55, 62, 227, 242, 252, 258, 259

environmental goods and services, 241-243, 262, 264

environmental responsibility, xvii, 140-143, 152, 155

Eora, xvi, xvii, 11, 17, 33, 63-67, 70, 170, 172-173, 208, 247

EUREAPA, 269-270

European regional policy, 221, 234, 238

Eurostat, 47, 54-55, 61, 220, 221, 226-228, 234, 238, 262, 273, 275

eutrophication, 56, 61, 159, 161, 168

EXIOPOL, xvii, 52-54, 56-58, 60, 61, 72, 157-159, 161, 207, 208, 221, 247, 273

extended input-output tables for the European Union, 221, 228

Extension Data, 12, 16

Food and Agriculture Organisation (FAO), 56

food security, 139

footprint analysis, 3, 125, 273

fossil fuels, 56, 85, 140, 195, 196, 198, 269

genuine saving, 230, 231

GLIO, xvi, xvii, 69, 70, 74-77, 180-181

Global Resource Accounting Model (GRAM), xvi, xvii, 79-80, 190-191

global supply chains, xiii, xvii, 7, 74, 136-137, 181, 182, 184, 187, 204, 207, 218, 242, 247, 260, 270

Global Warming Potential (GWP), 56

Green Growth, xvii, 207, 240-243, 245-248

Green water, 115, 118, 119

greenhouse gas, xvii, 12, 57, 79, 85, 86, 125, 129, 157-159, 161, 168, 172, 180, 191, 207, 241, 244, 246-248, 260, 262-263, 268

GTAP, xvi, xvii, 14, 16, 21-31, 63, 72, 125-127, 136-137, 267, 27

hybrid-LCA, 8

IEA, 56, 84, 86, 207

income responsibility, xvii, 141-142, 146, 150, 152-155

Industry-technology assumption, 58

input-output data, xvii, 27, 80, 100, 115, 226, 220-223, 230, 234, 242

input-output framework, 221, 222, 229, 234, 251, 259

input-output tables for the European Union, 221, 226, 228, 234, 238

intermediate demands, 12, 13

intermediate input shares, 81

interregional, 3, 4, 33-35, 41, 49, 115-117, 142

IO-LCA, 181, 182, 186, 187

IOT, 49, 52-54, 58-62, 255, 258, 259, 262, 264

IRIO, 5, 6

Japanese Products, 181, 183, 184, 185

Kyoto mechanisms, 134

land appropriation, 137-139

Land footprint, 61, 128

land use, xvii, 28, 50, 56-57, 61, 125-126, 128-134, 136-137, 140, 158, 161, 168, 260, 268, 273

land-intensive product, 136, 138

land-use intensities, 126, 131
life cycle assessment (LCA), 1, 180, 242
lifestyles, xvii, 89, 136, 137, 155
Low Emission Development, 245, 246, 247
Low-Carbon Model Town, 241, 245
Market Exchange rates, 16, 55, 83, 173
metals and industrial minerals, 85, 195-198
Monte-Carlo techniques, 15, 16
MRIO-based accounting, 244
multipliers, 15, 102, 108, 160-162, 262
Multi-regional Environmentally Extended Supply and Use Tables (MR EE SUT), 52
NAMEA, 53
national accounting, 231
non-survey method, 4, 5, 6
OECD countries, 79, 85, 141, 193, 195, 196, 198, 199, 204, 207-208
OECD input-output tables, 81, 83
OECD STAN, 83
optimisation, 17, 134
Ozone Depletion Potential, 56
Photochemical Oxidant Creation Potential (POCP), 56
policy applications, 17, 170, 208, 267
policy making, xvi, xvii, 192, 209, 210, 241, 274
problem shifting, 161, 169
process-based accounting, 264
producer responsibility, xvii, 63, 113, 141-142, 145, 152-155
production-based accounting, 85, 155
Proportionality Assumption, 47
purchaser prices, 54
Purchasing Price Parity, 16
RAS, 14, 83
Re-exports, 12, 209
regional integration, xvii, 220, 238
region-sector combination, 128

Resource Accounting, xvi, xvii, 2, 79, 80, 190-191
Rest of World, 11, 57, 59, 62, 192, 195, 272
risk, 10, 16, 17, 30, 205, 207
scenario, 85, 121, 154, 244, 270, 274
sector classification, 8, 13, 36, 40, 66, 149, 173
SEEA, xviii, 250-253, 258, 261, 264, 265
shared producer and consumer responsibility, 142, 147-151, 154
shared responsibility, 150, 154, 199
single-region input-output (SRIO), 3
Social Accounting Matrix (SAM), 25
social impacts, 30, 95
Spain, xvii, 44, 50, 113, 114, 117-118, 120
standard deviation, 17, 64, 67
Stockholm Environment Institute, 272
Structural Decomposition, xvii, 7, 170-172
structural path analysis (SPA), 7, 105
supply and use tables, 15, 43, 45, 52, 54, 220-222, 226-227, 234, 237-238, 251-253, 255
supply chain, xiii, xvii, 1-3, 7, 64-66, 69-70, 93, 103, 120, 125-127, 129-130, 134, 141, 145-150, 155, 158, 160, 180-181, 183, 186-188, 212-213, 241, 242, 244, 246-248, 251, 260, 265, 269, 274
Sustainable Consumption and Production (SCP), 271
sustainable development, xvii, 169, 207, 220-221, 230, 232-234, 238, 240, 242-243, 265, 270
The Institute of Developing Economies (IDE-JETRO), 33
TIIO, 34, 36, 41
Total Material Requirements (TMR), 273
trade barriers, 216, 218

trade coefficient, 4, 5, 208, 210
trade data, 5, 6, 12, 13, 26-29, 44,
    45, 67, 80, 83, 85, 115
Trade Facilitation, 217
trade in tasks, 211-212, 214, 217
trade in value added, xvii, 206, 211,
    212, 214, 216
trade policy, 205, 211, 216-218, 241,
    247
trade-off, xvi, 158, 168, 242, 249,
    262
transaction costs, 209, 216
UK Government, 271-273
UN COMTRADE, 26, 47, 57, 210
uncertainty, 10-17, 64-65, 67, 77,
    110, 218
valuation layers, 54
value added, xvii, 3-7, 11, 13-15, 30,
    44, 46, 49, 53, 81-83, 86, 130-
    132, 142-143, 149, 153, 159, 160,
    206, 211-216, 218, 226, 237-238,
    255, 258, 262
variability, 10, 11, 13-14
Virtual Water, 113, 121, 269
Waste input output, 95
Waste Resource Action Programme
    (WRAP), 272
water availability, 114, 118, 120
water consumptive use, 115, 119
water footprint, 64, 113, 114, 117,
    262, 269, 270, 273

www.ingramcontent.com/pod-product-compliance
Lightning Source LLC
Chambersburg PA
CBHW060028030426
42334CB00019B/2231